Communication for Continuous Improvement Projects

Industrial Innovation Series

Series Editor
Adedeji B. Badiru
Department of Systems and Engineering Management
Air Force Institute of Technology (AFIT) – Dayton, Ohio

PUBLISHED TITLES

Carbon Footprint Analysis: Concepts, Methods, Implementation, and Case Studies, *Matthew John Franchetti & Defne Apul*

Communication for Continuous Improvement Projects, *Tina Agustiady*

Computational Economic Analysis for Engineering and Industry, *Adedeji B. Badiru & Olufemi A. Omitaomu*

Conveyors: Applications, Selection, and Integration, *Patrick M. McGuire*

Global Engineering: Design, Decision Making, and Communication, *Carlos Acosta, V. Jorge Leon, Charles Conrad, & Cesar O. Malave*

Handbook of Emergency Response: A Human Factors and Systems Engineering Approach, *Adedeji B. Badiru & LeeAnn Racz*

Handbook of Industrial Engineering Equations, Formulas, and Calculations, *Adedeji B. Badiru & Olufemi A. Omitaomu*

Handbook of Industrial and Systems Engineering, Second Edition *Adedeji B. Badiru*

Handbook of Military Industrial Engineering, *Adedeji B.Badiru & Marlin U. Thomas*

Industrial Control Systems: Mathematical and Statistical Models and Techniques, *Adedeji B. Badiru, Oye Ibidapo-Obe, & Babatunde J. Ayeni*

Industrial Project Management: Concepts, Tools, and Techniques, *Adedeji B. Badiru, Abidemi Badiru, & Adetokunboh Badiru*

Inventory Management: Non-Classical Views, *Mohamad Y. Jaber*

Kansei Engineering - 2 volume set

- Innovations of Kansei Engineering, *Mitsuo Nagamachi & Anitawati Mohd Lokman*
- Kansei/Affective Engineering, *Mitsuo Nagamachi*

Knowledge Discovery from Sensor Data, *Auroop R. Ganguly, João Gama, Olufemi A. Omitaomu, Mohamed Medhat Gaber, & Ranga Raju Vatsavai*

Learning Curves: Theory, Models, and Applications, *Mohamad Y. Jaber*

Modern Construction: Lean Project Delivery and Integrated Practices, *Lincoln Harding Forbes & Syed M. Ahmed*

Moving from Project Management to Project Leadership: A Practical Guide to Leading Groups, *R. Camper Bull*

Project Management: Systems, Principles, and Applications, *Adedeji B. Badiru*

Project Management for the Oil and Gas Industry: A World System Approach, *Adedeji B. Badiru & Samuel O. Osisanya*

Quality Management in Construction Projects, *Abdul Razzak Rumane*

Quality Tools for Managing Construction Projects, *Abdul Razzak Rumane*

Social Responsibility: Failure Mode Effects and Analysis, *Holly Alison Duckworth & Rosemond Ann Moore*

Statistical Techniques for Project Control, *Adedeji B. Badiru & Tina Agustiady*

STEP Project Management: Guide for Science, Technology, and Engineering Projects, *Adedeji B. Badiru*

Systems Thinking: Coping with 21st Century Problems, *John Turner Boardman & Brian J. Sauser*

Techonomics: The Theory of Industrial Evolution, *H. Lee Martin*

Triple C Model of Project Management: Communication, Cooperation, Coordination, *Adedeji B. Badiru*

FORTHCOMING TITLES

Cellular Manufacturing: Mitigating Risk and Uncertainty, *John X. Wang*

Essentials of Engineering Leadership and Innovation, *Pamela McCauley-Bush & Lesia L. Crumpton-Young*

Sustainability: Utilizing Lean Six Sigma Techniques, *Tina Agustiady & Adedeji B. Badiru*

Technology Transfer and Commercialization of Environmental Remediation Technology, *Mark N. Goltz*

Communication for Continuous Improvement Projects

TINA KANTI
AGUSTIADY

CRC Press
Taylor & Francis Group
Boca Raton London New York

CRC Press is an imprint of the
Taylor & Francis Group, an **informa** business

CRC Press
Taylor & Francis Group
6000 Broken Sound Parkway NW, Suite 300
Boca Raton, FL 33487-2742

Printed on acid-free paper
Version Date: 20130819

International Standard Book Number-13: 978-1-4665-7775-6 (Hardback)

Library of Congress Cataloging-in-Publication Data

Agustiady, Tina.
Communication for continuous improvement projects / Tina Agustiady.
pages cm -- (Industrial innovation series ; 29)
Includes bibliographical references and index.
ISBN 978-1-4665-7775-6 (hardback)
1. Quality control. 2. Communication in management. 3. Industrial management. I. Title.

TS156.A35 2014
658.4'013--dc23 2013032498

Visit the Taylor & Francis Web site at
http://www.taylorandfrancis.com

and the CRC Press Web site at
http://www.crcpress.com

To my first born child, Arie Agustiady. I stayed motivated each day because of our love for one another.

To my dear husband, Andry, you encouraged me each step of the way and make me a stronger woman!

Contents

Preface

> Believe in yourself! Have faith in your abilities! Without a humble but reasonable confidence in your own powers you cannot be successful or happy.
>
> —Norman Vincent Peale

Benefits are always needed in companies in order to be successful and profitable, and for employees to justify their self-worth. Although Continuous Improvement techniques are sought after, the implementation of the techniques becomes difficult and challenging to sustain. The proper tools and communication within organizations are key to making decisions and having successful results. Continuous Improvement requires project management and accountability along with the proper tools and communication to be effective. To sustain results, relying on an individual without the proper methodology is simply not enough. True statistical techniques need to be implemented to help make each industry the best in what they do. Lean and Six Sigma are important tools that are here to stay. The tools from Lean and Six Sigma are not only statistics, but show data-driven decisions that will implement results based on facts. Communicating these tools is the most difficult part of using the tools. This book will make Lean Six Sigma such an easy methodology that everyone will want to jump on board. This book will help companies become more marketable and profitable. The economy needs an art of effective communication utilizing the right set of tools and techniques in an easy, defined manner.

Continuous Improvement is the ongoing improvement of any efforts, which are mainly used in business or manufacturing practices. Customer-focused results are sought after with the aim to continuously make customers happy and to have a competitive business. Communicating Continuous Improvement is the most important part of any improvement process due to the need for cross-functional intelligence of employees.

Continuous Improvement is a growing topic now, and almost all industries are aspiring to the financial benefits. Utilizing Continuous Improvement, however, is not clearly defined. This book will help industries and engineers learn how to be a change agent, gaining and maintaining positive results. The book will use specific tools utilizing Lean Six Sigma, and then demonstrate the communication aspect of how to use the tools and how to be an effective leader.

Manufacturing companies are working endlessly to make improvements, which are difficult to implement and even more difficult to sustain. The key is to communicate why Continuous Improvement methodologies are being used and how to sustain the improvements. Most important, employees want to be empowered and believe that the tools they are utilizing are needed.

This book will show how to be an effective change agent by using tools that make sense while being competitive in the business market. Employees will learn to be successful and will be able to show quantifiable results. The proper tools, communication, and management make the methodologies of Lean Six Sigma work. This book also includes a Continuous Improvement Toolkit (Chapter 11) that is an easy reference for which tools to use, and when and how to effectively teach the tools to employees who are not necessarily engineers. The implementation of the actual tools is also taught in this book. Result-driven decisions can be made from the methodologies used in this book, making processes quantifiably better with sustainable results. There will no longer be guesswork after using this extensive and informative book showing the art of Continuous Improvement through communication.

The Continuous Improvement Toolkit consists of the following topics, which will be presented using real-life examples:

5S
7/8 Wastes
Kaizen
Fishbone Diagrams
Root Cause Analysis
Process Mapping
Financial Justification
One Point Lessons
Value Stream Mapping
Plan–Do–Check–Act
Poka Yokes
Kanbans
Pull and Push Flows
Visual Management
Cellular Processing
7/8 Wastes
Spaghetti Diagrams
Histograms
Pareto Charts
Capability Analysis
Control Charts
Defects per Million Opportunities
Project Charters
Supplier–Input–Process–Output–Customer (SIPOC)

Kano Model
Critical to Quality (CTQ)
Affinity Diagram
Measurement Systems Analysis
Gage R&R
Process Capabilities
Variation
Graphical Analysis
Cause and Effect Diagram
Failure Mode and Effect Analysis (FMEA)
Hypothesis Testing
Analysis of Variance (ANOVA)
Correlation
Simple Linear Regression
Theory of Constraints
Single Minute Exchange of Dies (SMED)
Total Productive Maintenance (TPM)
Design for Six Sigma (DFSS)
Quality Function Deployment (QFD)
Design of Experiments (DOE)
Mood's Median Test
Control Plans

Author

Tina Agustiady is a certified Six Sigma Master Black Belt and Continuous Improvement Leader at BASF. Agustiady serves as a strategic change agent, infusing the use of Lean Six Sigma throughout the organization as a key member of the Site Leadership team. Agustiady improves cost, quality, and delivery at BASF through her use of Lean and Six Sigma tools, while demonstrating the improvements through a simplification process. Agustiady has led many Kaizen, 5S, and Root Cause Analysis events throughout her career in the health care, food, and chemical industries.

Agustiady has conducted training and improvement programs using Six Sigma for the baking industry at Dawn Foods. Prior to Dawn Foods, she worked at Nestlé Prepared Foods as a Six Sigma product and process design specialist responsible for driving optimum fit of product design and current manufacturing process capability, and reducing total manufacturing cost and consumer complaints.

Agustiady received a B.S. in industrial and manufacturing systems engineering from Ohio University. She earned her Black Belt and Master Black Belt certifications at Clemson University (South Carolina).

Agustiady is also the Institute of Industrial Engineers (IIE) Lean Division President and served as a board director and chairman for the IIE annual conferences and Lean Six Sigma conferences. She is an editor for the *International Journal of Six Sigma and Competitive Advantage.*

Agustiady is an instructor who facilitates and certifies students for Lean and Six Sigma for IIE and Six Sigma Digest. She spends time writing for journals and books while presenting for key conferences. Agustiady is also the coauthor of *Statistical Techniques for Project Control and Sustainability: Utilizing Lean Six Sigma Techniques* (Taylor & Francis/CRC Press, 2012).

1

Effective Communication

> When people talk, listen completely. Most people never listen.
>
> —Ernest Hemingway

Communication begins with listening and listening well. Ensuring what someone has to say and understanding what they said is key to having an effective response (Figure 1.1). In order to respond in an engaging fashion, the recipient must know that you truly heard what they had to say and are not simply speaking what you wanted to say. Reiterating what the person has to say before responding shows the person that you truly heard them and are paying attention to them. People in general like attention and being attended to. When you respond in a fashion that their words matter and make a difference, you impact that person. Asking questions about the comments made also show the person that you are looking to have a conversation with them. Ensuring the questions do not have a negative connotation shows the positivity that the conversation can bring and will make the person want to engage back into the conversation. A person's tone can tell a lot about the conversation as well. It is important to see that someone is interested in having a conversation before continuing in the conversation. If the person is distracted or busy, the conversation will end up being meaningless and unproductive. If the conversation is important and will take more time than a few moments, it is important to ensure etiquette and ask the person for their time before starting a conversation. It is not impolite to tell someone that you do not have the time at the moment, as long as another time to have the conversation is given. Letting the person know that the conversation is important to you is also important so the recipient does not feel slighted.

When the conversation does happen, if something is misunderstood, ensure it is clarified then and not later. Confusion in communication is as bad as no communication. Assuming meanings of the text without clarification can lead to problems as well.

When listening to a person, utilize facts versus egos, opinions, and emotions. People may come to you with a great deal of emotion. When facing these people, allow them to collect themselves, but ask for the facts. Egos will be played in to these emotions by people speaking about them being better than others or people stating others' incompetency. Again, gain the facts out of the conversation and discuss the situation on a factual basis. It is important to never give one's opinions back about other people and only staying

FIGURE 1.1
Engaging effective responses.

with the facts. It is important not to add to people's egos but to stay consistent with stories and information.

Innovation helps deviate from the status quo and helps conversations stay lively. Innovation during communication is driven by asking questions out of the norm. Questions that require thought and consideration drive innovative answers and drive creative thinking. "Outside the box thinking" is when ideas are thought of as more innovative than normal and would not be thought of as the common answer or thought. This type of thinking should be utilized when communicating to be innovative so that people are more interested and want to think outside the box themselves. Innovative thinking also challenges people to want to be different and try to change for the better. The innovative thinkers and communicators are the people who change conversations from being general to being exciting. This type of conversation always leads to people wanting to continue the conversation or come back to discuss other items.

The same methodology goes for communication within projects. Clear communication of project requirements, constraints, and available resources are necessary for projects to have successful completions and results. The success of projects depends on good levels of communication, cooperation, and coordination.

Projects sometimes fail in spite of concerted efforts to make improvements or have successful projects. Understanding the most common reasons projects fail is important so the burden is overcome. The main reasons that projects fail include the following (also see Figure 1.2):

- Lack of communication
- Lack of cooperation
- Lack of coordination
- Diminished interest
- Diminished resources
- Change of objectives
- Change of stakeholders
- Change of priority
- Change of perspective

- Change of ownership
- Change of scope
- Budget cut
- Shift in milestones
- New technology
- New personnel
- Lack of training
- New unaligned capabilities
- Market shift
- Change of management philosophy
- Manager moves on
- Depletion of project goodwill
- Lack of user involvement
- Three strikes followed by an out (too many mistakes)

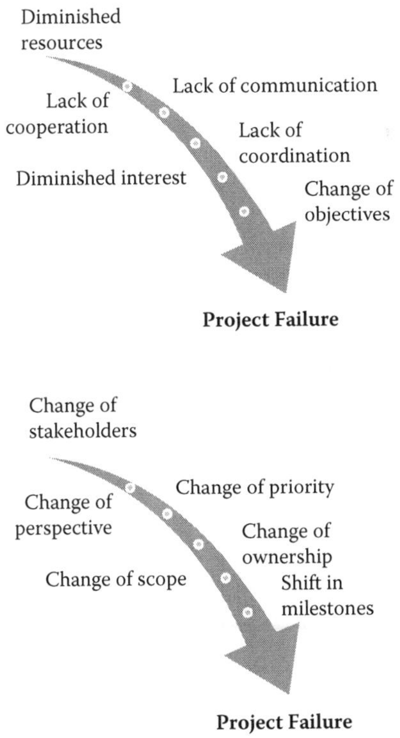

FIGURE 1.2
Main causes of project failure.

To minimize project failure rates, communication must increase and projects must be organized. Project organization specifies how to integrate the functions of the personnel involved in a project. Organizing is usually done concurrently with project planning. Concerted supervision and guidance is an important aspect of project organization and is a crucial aspect of the management function. Management requires skillful managers who can interact with subordinates effectively via good communication and motivational techniques. A good project manager will facilitate project success by directing his or her staff through proper task assignments toward the project goal. Workers perform better when they have clearly defined expectations. They need to know how their job functions contribute to the overall goals of the project. Workers should be given some flexibility for self-direction in performing their functions. The manager should recognize individual worker needs and limitations. Directing a project requires motivating, supervising, and delegating skills. The project manager must have prompt access to individual activity data and overall status of the project. Routinely clarifying project objectives will ensure that employees are aligned. Clearly defining needs and resources for the project prior to the project beginning will help the performance of the project. Once the project begins, details must be gathered in order for the tasks to be performed properly and analysis on the project to be completed. A file or visual representation of the project should be completed for documentation purposes.

Leadership is important for the projects to succeed. Leadership is a key aspect of management. Leadership requires self-direction to inspire, set goals, supervise, and empower relationships. An effective leader must have extraordinary communication skills. Communicating business plans, strategies, and objectives in a detailed manner conveys the top-down vision of a business. A leader must be responsible and never pass blame onto others. He or she should find the root cause of a problem and analyze the reasons for failures. Delegating responsibility for tasks also shows leadership capabilities. After a task is communicated, it is important for a leader to ensure that the employees understood the task. The "five why" tool is useful at this point. Asking why five times drills in the true understanding of a task (Figure 1.3).

The answers reveal knowledge of the process, systems, and business. Communication is a main aspect of management. Two-way communication is required. Employees must be able to confide in management about difficult situations; management must be able to explain expectations and timelines. Communication can boost positive morale throughout an organization. Positive reinforcement helps employees maintain structure and commitment, and builds self-esteem. Positive attitudes are contagious.

Dr. W. E. Deming proposed the "Theory of Profound Knowledge" and said it was important for any business that desired to be competitive. According to Deming, for employees or a business to be successful, the system depends

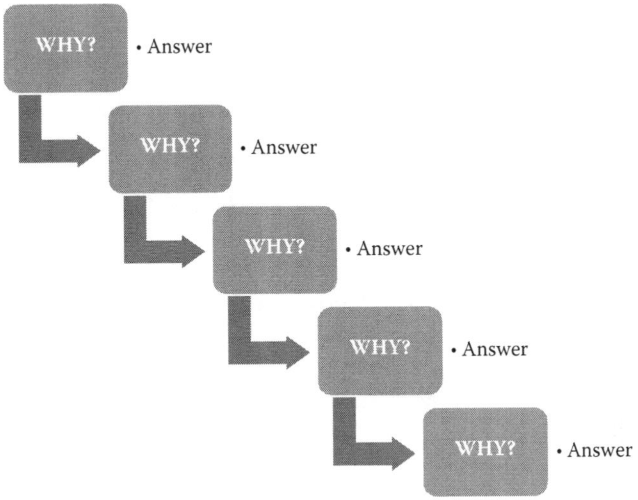

FIGURE 1.3
Ask why five times.

on management's ability to balance each component to optimize every system. The four categories of profound knowledge are:

- Systems (processes) (Figure 1.4)
- Statistics (variation) (Figure 1.5)
- Psychology (motivation) (Figure 1.6)
- Knowledge (Figure 1.7)

The knowledge of systems involves having outside management improve a system because a system has no knowledge. Quality management is the key for a system to ensure the end goal is met. If customers are not happy with the existing system, they can easily go to a competitor. Constantly improving a system by understanding its components will demonstrate Continuous Improvement.

Statistics knowledge reveals details of variations. Variation happens over time and should be predictable. Statistics shows different patterns and types of variations. The end goal is to reduce variation as much as possible by identifying common cause and special cause variations. Management should be aware of common cause variations and when they may occur.

The knowledge of psychology is important for leaders because they want to motivate employees and all employees want to be motivated. If employees do good work, they want to be recognized. It is important for a leader to keep employees from fearing that they will lose their jobs, their value, and their recognition. Leaders must demonstrate trust in their employees and other leaders and must ensure that employees trust one another.

FIGURE 1.4
Systems (processes).

The knowledge of knowledge consists of the facts and information that employees gain over time after they learn a job or a system. The knowledge can also be taught by others or through formal training. Knowledge should be transferred; no single person should hold all the knowledge. Many people should understand a system so that if a key stakeholder is absent, an entire process does not stop, fail, or incur more variation.

Deming's philosophy utilized (a) versus (b) comparisons:

(a) When people and organizations focus primarily on quality, defined by

$$\text{Quality} = \text{Results of work efforts/Total costs}$$

quality tends to increase and costs fall over time.

FIGURE 1.5
Statistics (variation).

FIGURE 1.6
Psychology (motivation).

(b) When people and organizations focus primarily on costs, costs tend to rise and quality declines over time. Deming developed a cycle for this principle from the work of Walter A. Shewhart, an American physicist, engineer, and statistician considered the "Father of Statistics." The cycle is called the Deming cycle or the Shewhart cycle and involves four steps: plan, do, check, act (PDCA). The PDCA steps for a project are planning the study or designing the experiment, performing the required steps, checking the results and testing the information, and acting on the data-driven results.

Engaging and inspiring employees and people is one of the most sought out ways to have effective communication. Adaptability toward different people and groups is important so that people are engaged in the topics covered. Adaptability is about learning about different people and understanding

FIGURE 1.7
Knowledge.

what motivates them. Creating energy around these people and the things that inspire them make the employees want to succeed. Some employees are self-driven and do not need motivation, but almost all employees want some type of incentive or want to be appreciated for their contributions. People enjoy either working independently or within groups. If employees like working independently, they most likely seek recognition for their efforts and want to be known to be in control and having power. Incorporating a teamwork setting helps people work better together because the communication is at a higher level. This type of setting also helps employees seek out relationships to build on. People also like being in charge, so giving people trust and communicating that trust allows people to be competent. Most people want to succeed, so giving them the trust to work on a project is normally motivation enough to work diligently and have good outcomes. It is important during this process to have status updates and be available for questions and comments. Also, it is important to support the person that the trust has been given to without just inputting one's own opinion on how it should be done. It is extremely important to not try and take the project back over if it is not going as planned. There are many reasons for different ideas, and without cross-functional teams and ideas there would be no growth in the business.

When giving others responsibility, it is important that all team members are aware of the mission and vision. If there was a strategy on how the work needed to be done, it is very important to be transparent. Transparency leads to people having as much knowledge as they possibly should to do a job well done. Transparency does not have to be about giving confidential information, but instead about providing the big picture on what is expected and desired and what the final anticipated outcome should be. It is also important that all people have the same vision and agree on what needs to be done and why. If people do not agree with the final outcome or game plan, they will be less likely to want to cooperate or participate. If there is disagreement in the vision, explain the reason the vision was given and the thought process behind it. Even if there is not 100% agreement on the vision, the transparency on why the vision was given will help show the goals are realistic and achievable. Being open to challenging the status quo at this point in time is important, but once the vision is completed, it is important to move forward and not take too many steps backward.

Fostering open communication should become a lifestyle versus a job. This like any core competency entails work and must be exercised daily in order to be effective. Becoming a good leader does not happen overnight. It requires support with clear expectations of oneself and from others. Ensuring people have the information they need to be effective and do the appropriate job is all part of communication. Once people realize the goal is common and the information is available, there will be fewer disagreements and the leadership qualities will be more transparent. The communication strategy should also be part of the overall strategy so that it is apparent that transparency is a key value and that the communication is coming from all levels. This methodology should be conveyed top down and bottom up. It is important that all people know what their communication strategy will be whether it is visual or verbal. The structure behind this plan will help communication in the long run. It is also important to make communication not just data collection, but a means of growing teams and individuals. This will help people want to have ownership in their area of expertise and encourage others to want to participate. If people do not want to participate, giving them positive reinforcement will help them overcome their feelings of exclusion. Gaining feedback on the present communication is important, but encourage employees not to speak hastily or give negative remarks. Encourage people to give positive reinforcement and suggestions versus bad news or insight on their communication style. It is important to have diversity in communication; this makes the communication more open and fun. It is also important to have any continuous negativity pointed out and addressed so others do not feel uncomfortable. If too much negative feedback is given in a public setting, the communication means can change to being not face to face. It is also good for people to give anonymous feedback so that they are more open. Again, inform the group that the feedback is taken generously and to avoid being hasty and condescending in the feedback; instead give feedback in a means that will help other people to grow.

Receiving feedback is very important for people to grow. Sometimes the feedback may be taken negatively, but it should not be. It is important to think about the suggestions and ways to change a habit based on the feedback. Thinking about the information before acting on it immediately will help as well. Showing people that the feedback was actually used will encourage them to provide feedback more often and to open up in terms of communication. If the feedback is not provided as desired, asking questions can help the process. The normal questions to ask are very basic, but provide good guidance:

- What should we be communicating?
- What are we not communicating?
- Are you missing any information that would help you to do your job?
- What do you need from me to be an effective leader?
- Are there any barriers?
- What motivates you?

Once these type of questions are asked, the answers should be shared once gathered in an organized fashion. It should be shown that particular changes will be made due to the feedback incurred by these questions. Showing the results of feedback will encourage more feedback in the long run.

Knowing what other people's knowledge areas are is also an important means of communication. This knowledge sharing area can help others go to the appropriate people who are the subject matter experts. It is important that there are cross-functional knowledge experts and who those people are. This should be as transparent as possible.

Finally, timely updates on progress is one of the most needed means of communication. These updates should be given periodically so that everyone is aware of what has already been done and what needs to be done so redundant work is not completed. This will also help with goal alignment and show the natural progression of success and the means to get to the end.

Confidence is required in leading and communicating effectively. It is important for other people to gain respect for you. Respect is earned, not given; a particular position does not entitle a person to be respected. Showing the reason the position was given to a particular person due to their knowledge and ability will help that respect to be given. Intimidating other employees will only make people lose respect. Letting people know your values and why you are making the decisions that you are will foster open minds and communication. Data-driven results are more important than egos, opinions, and emotions. Advising people of the data behind decision making makes them more confident in believing the goals and strategies. If opinions are given on why a topic is to be completed, it is important in letting others know the reasoning behind it. Giving past experiences to convince others is a methodology that should be used to show the transparency in the decision making. The end goal is to ensure the effectiveness of the communication is occurring. It is clear when communication is effective because fewer mistakes will occur and the end goal is achievable while innovative thinking occurs concurrently.

Workplace Communication

Communication in the workplace can be your greatest ally when used effectively. It can help to avoid confusion, set clear goals and deadlines, create healthy work relationships, set yourself apart as a leader, and keep everyone on the team on the same wavelength. However, on the other hand, poor communication can sabotage even the most talented team of individuals and is an easy way to set up eventual failure. There are several types of communication, including written, verbal, nonverbal, and electronic (Figure 1.8).

Verbal communication is the most common form of communication, but can also be the hardest to master. When verbally communicating in the workplace, you must be clear, direct, and detailed so that no questions are left unanswered or assumptions made do not eventually create mistakes. An overview of a path forward and action items after a discussion is often

FIGURE 1.8
Workplace communication.

a good way to review the details of a conversation or meeting while also allowing everyone in the group to clear up any misconceptions or confusion. A clear path of communication between all parties must be present—one overbearing individual dominating a conversation could influence others with good ideas to hold their tongue. Receptive listening is just as important as effective speaking and these points all make up what is good verbal communication.

Nonverbal communication is part of the workplace life that can make or break relationships. Nonverbal cues that one is not even conscious of making can hinder or even break lines of communication. Activities like rolling of the eyes, lack of attentiveness, and ignoring others can hurt the overall effectiveness of open communication and even cause others to avoid or shutdown communication with you. More simple positive cues such as eye contact, nodding of the head, and being attentive can help to show the speaker that you are an eager and willing listener, and can keep lines of communication open and clear.

Written as well as electronic communication can be just as important as direct verbal communication. How reports, memos, text messages, and especially e-mails are written can help catch someone up to speed on a subject or confuse them further, depending on the quality and detail. For instance, if a report or memo is being sent to a group of people, that particular document should be written specifically for that audience and not for yourself. Assumptions based on knowledge can lead to entire documents being misunderstood or ignored because of a lack of understanding. Sufficient background information as well as an explanation of technical terms can help everyone stay on the same page and help cultivate a lively and effective conversation. E-mails and text messages, while often used less formally, are very important in the workplace, too. Keeping the right people on e-mail chains, and approaching subjects the right way will keep coworkers open to communication and apprised of all important details. E-mails should never be responded to in an accusatory way, and should only contain fact rather than hearsay or speculation. Such things can lead to confusion and misinformation that can often lead teams down the wrong path. Rather, e-mails with clear details, background information, the correct audience, and breakdown of specific tasks are strong tools to keep lines of effective communication open.

The process of exchanging information in an efficient manner will make communication effective. Information is conveyed as words, tone of voice, and body language. Studies have shown that words account for 7% of the information communicated. Vocal tone accounts for 55% and body language accounts for 38%. To be effective communicators, people must be aware of these types of communication methods, how to use them effectively, and barriers that negate the communications process.

Effective communication consists of conveying the proper message when needed and especially when questioned, giving data or rationale behind the information given, and repeating the information in a consistent manner. Effective communication is based on direct data and should not be based on opinions. The communication must be clear and concise with proper terminology, tone, and focus.

References

Agustiady, Tina, and Adedeji B. Badiru. 2012. *Statistical Techniques for Project Control.* Boca Raton, FL: Taylor & Francis/CRC Press.

Agustiady, Tina, and Adedeji B. Badiru. 2012. *Sustainability: Utilizing Lean Six Sigma Techniques.* Boca Raton, FL: Taylor & Francis/CRC Press.

United States Coast Guard. Team Coordination Training. http://www.uscg.mil/auxiliary/training/tct.pdf.

2

Best in Class Practices

Badiru's rule: The grass is always greener where you most need it to be dead.
—Adedeji B. Badiru

Think about the first immaculate piece of technology you worked with. Shortly after, weren't there very similar pieces of that same technology, but maybe even a little bit better? How did that happen? Utilizing Best in Class (BIC) practices, better and better items are made. A person takes the BIC material, determines exactly how to produce it, and then essentially makes it better. If the first BIC product is not replicated, there would be a monopoly of companies being the only company in the business making that particular product. Each company now takes the BIC products and continuously improves them in order to stay competitive in the marketplace. If each company does not continuously improve or is not innovative, the competition will take the business and the customers will follow quickly.

Project management should be an enterprise-wide, systems-based endeavor. Enterprise-wide project management is the application of project management techniques and practices across the full scope of an enterprise. This concept is also referred to as management by project (MBP), an approach that employs project management techniques in various functions in an organization. MBP recommends pursuing endeavors as project-oriented activities. It is an effective way to conduct any business activity. It represents a disciplined approach that defines each work assignment as a project. Under MBP, every undertaking is viewed as a project that must be managed just like a traditional project. The characteristics required of each project so defined are:

- Identified scope and goal
- Desired completion time
- Availability of resources
- Defined performance measure
- Measurement scale for review of work

An MBP approach to operations helps identify unique entities within functional requirements. This identification helps determine where functions overlap and how they are interrelated, thus paving the way for better planning, scheduling, and control. Enterprise-wide project management facilitates a unified view of organizational goals and provides a way for project teams to use information generated by other departments to carry out their functions. The use of project management continues to grow rapidly. The need to develop effective management tools increases with the increasing complexity of new technologies and processes.

The life cycle of a new product to be introduced into a competitive market is a good example of a complex process that must be managed with integrative project management approaches. The product will encounter management functions as it goes from one stage to the next. Project management will be needed throughout the design and production stages, and will be needed in developing marketing, transportation, and delivery strategies for the product. When the product finally reaches customers, project management will be needed to integrate its use with those of other products within customer organizations.

The need for a project management approach is established by the fact that a project will always tend to increase in size even if its scope is narrowing. The following four literary laws are applicable to any project environment:

Parkinson's law—Work expands to fill the available time or space.

Peter's principle—People rise to the level of their incompetence.

Murphy's law—Whatever can go wrong will.

Badiru's rule—The grass is always greener where you most need it to be dead.

An integrated systems project management approach can help diminish the adverse impacts of these laws through good project planning, organizing, scheduling, and control.

How to Stay Best in Class

A best practice is a technique or methodology that, through experience and research, has proven to reliably lead to a desired result. A commitment to using the best practices in any field is a commitment to using all the knowledge and technology at one's disposal to ensure success.

In today's economic market, if you do what you have always done, you will get what you always got. Customers may presently be happy with the product or service, but once they find the innovative new product or service, they automatically are drawn to it no matter the loyalty of the customer. Adopting philosophies and methodologies that make the business BIC is what makes a competitive business.

Utilizing new technology in the business will help the business grow by being more efficient and innovative. The systems being used help the business grow and the benefits of the new technologies attract new customers.

Providing communication with key customers and suppliers is important so they are in tune with changes and know the want for the business to keep growing and the desire to be BIC. This helps relationships stay healthy for both ends, and a false representation of what is to be produced is not evident.

The business should use a total cost approach versus a price of product approach. This includes all aspects of the business. Therefore, labor and materials are not the only overhead, but the entire supply chain is looked upon. What sourcing is the cheapest, what methods of transportations can be used to cut costs, and what bulk quantities need to be made to not overproduce and keep the customers happy? Operating, training, maintenance, warehousing, environmental, and quality costs should be under this umbrella as well. Working with suppliers, it should be asked how to reduce costs for both businesses in order to stay BIC in the overall business that is being conducted. Controlling processes is an essential part to being BIC because the process is standardized and validated for profit. Risk should be sought after to be eliminated so no unexpected events occur that increase costs more than normal operating practices. Identifying and eliminating risk should include:

- Identifying areas of risk
- Determining probability, occurrence, and detection of risk
- Determining costs of risk
- Prioritizing risk for monitoring and prevention

Best practices are known to spread through companies and fields of interest after a success occurs. Skills and knowledge are key practices for having BIC techniques. A BIC is also a technique that is considered consistently superior in the results produced, so benchmarking techniques are utilized to mimic leading practice results.

Benchmarking endeavors involve strategic and analytical planning. The competitive strategy is simply to acquire knowledge of the BIC process or product by comparing the processes and understanding the methodology. Once the process is understood and compared for differences, the process should be made more efficient, with a similar methodology without copying the exact product. If any items do not make sense of the processes, they should be eliminated and made into better processes so the new product or service is the new BIC technique.

There are four main types of benchmarking practices:

1. Internal Benchmarking
2. External Benchmarking
3. Functional Benchmarking
4. BIC Practices

Internal benchmarking seeks current practices within the business and improves its own functions and processes. Processes that are better than others are looked at for the BIC approaches and those approaches are used for new processes. External benchmarking is a comparison of a business that makes similar products. This benchmarking is used not to copy the other product or service, but instead to use the innovative ideas that are attractive to customers to improve one's own product or service. External benchmarking is normally completed with a competitor. Functional benchmarking focuses on the same functions within the same industries or businesses to be competitive without severe analytical or strategic skills. It takes high-level BIC processes and gives customers the same advantage. Generic benchmarking takes BIC practices across any organization whether the same or not, and utilizes the attractive aspects to make their own businesses better.

Strategic planning utilizes project management techniques to dissect technology in order to make a better business plan. It utilizes goals, visions, and strategies of the business to ensure alignment is present of being the best in the business. Analyses must be present during strategic planning so that the entire process is mapped without jumping to improvement measures before the definition and measurements have taken place.

Benchmarking and BIC practices come from communication from the customer. The customer is the number one priority, so feedback from the customer must be obtained in order to understand what it is the customer wants. If a new label is put on yearly to attract the customer, the customer may be attracted, but truly not care about the label and care more about the quality of the product. A voice of the customer analysis should be conducted to interpret benchmarking the right opportunities.

The speed in which the BIC practices take place is important because customers do not want to wait to receive their items. They want rush delivery on the best quality service they can find at a competitive price. If they do not get what they desire, they will go to another business without even thinking twice about the business that is making them wait (Figure 2.1).

Communication of findings must be shared in order for the strategy to continue. Working outside the office helps with strategy because it is seeking other endeavors or products to innovate. Most of the time ideas spark from brainstorming other products or services to find new products or services that are very different in nature, but represent what the customer is looking for.

The Best Manufacturing Practices (BMP) Program encourages companies to operate at a higher level of efficiency and become more competitive through the implementation and sharing of best practices. The BMP Program was established to enable the U.S. defense industry and the entire U.S. industrial base to improve product price, quality, and delivery. These type of practices can be found in any country and is encouraged as a means to have competitive advantages for companies.

Finding the gap between your particular business sector and the competitor is a beginning means to understand the amount of work it will take to

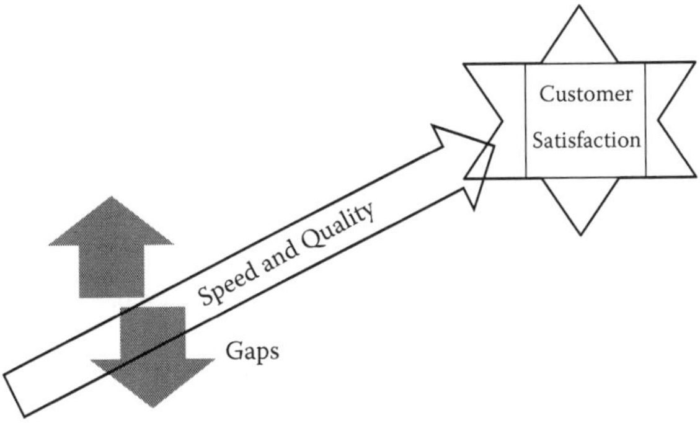

FIGURE 2.1
Speed and quality versus gaps.

get to the BIC process. Understanding key business metrics and standard operating procedures is also one of the first steps in finding ways to get to BIC practices. Understanding where to begin focus improvement projects is a means to improving so that the project stays in scope and is actually attainable. As practice of a business is engaged, performance automatically increases (Figure 2.2).

The operational performance of a business is based on the solid performance of the business and the structure at which it operates. If this strategy is analyzed properly, the business can be Best in Class at any operation it desires (Figure 2.3). To ensure the business is operating at a model desirable to the company, scorecards for BIC practices can be conducted. These include evaluation of the following key metrics:

- Quality Practices
- Quality Performance
- Lean Manufacturing Practices
- Cost Metrics
- Supply Chain Performance

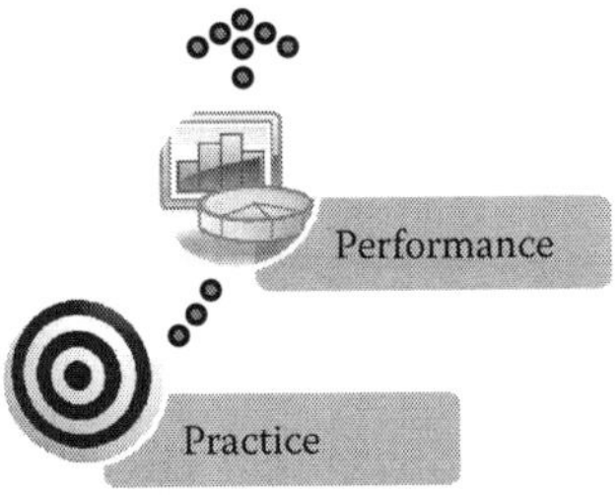

FIGURE 2.2
Practice versus performance.

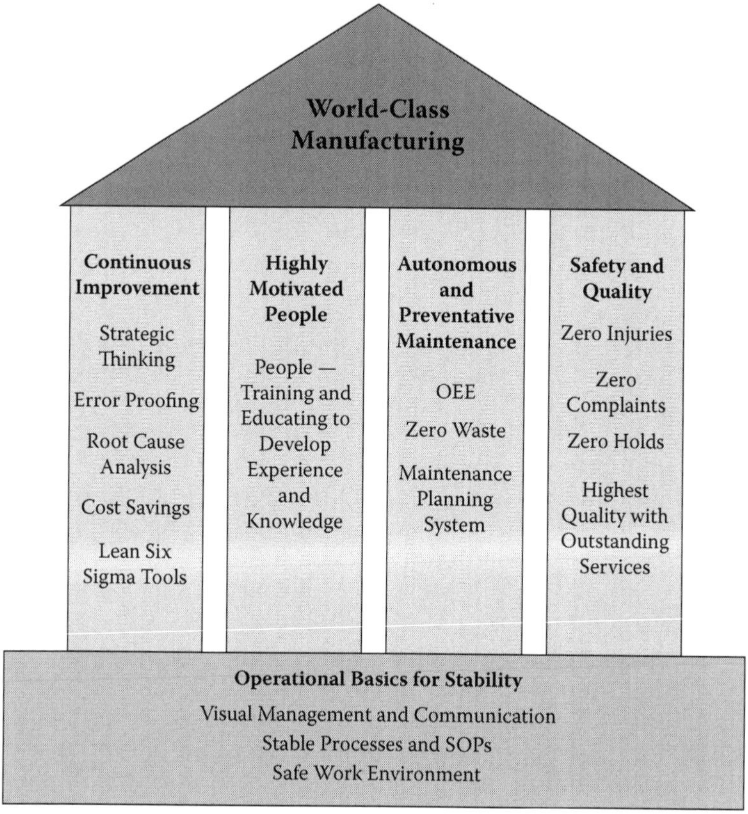

FIGURE 2.3
Operational stability model.

- Customer Satisfaction
- Leadership and Communication Skills

The amount of resources put into these metrics will ensure the business stays competitive in the marketplace.

The company view should include the SIPOC model (Figure 2.4), which shows

- Suppliers
- Inputs
- Processes
- Outputs
- Customers

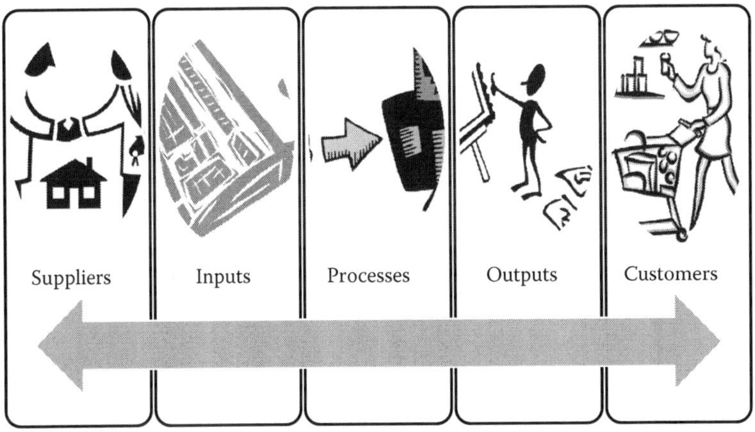

FIGURE 2.4
SIPOC model.

The view involves understanding the processes first and foremost. Processes are management's responsibility; management must understand all processes and the rationale behind the processes. The inputs and outputs of the processes are looked upon next to ensure there are no gaps and the correct output is being produced. Finally, having supplier and customer relations is an essential part to having BIC practices because the alignment must be present in order for the company to succeed and improve.

References

Best Manufacturing Practices Center of Excellence. Best Practices Surveys: Overview. http://www.bmpcoe.org/bestpractices/.

Tech Target. Definition: Best Practice. http://searchsoftwarequality.techtarget.com/definition/best-practice.

3

Maintaining Sustainability

> People "over-produce" pollution because they are not paying for the costs of dealing with it.
>
> —Ha-Joon Chang

Sustainability

Sustainability is the human preservation of the environment, whether personally, socially, or economically through responsible management of resources, education, and continuous process improvement (Figure 3.1). Being sustainable corporate citizens will increase sales and profitability by reducing costs and having a competitive advantage.

The management discipline organizes and allocates efforts and resources across the broad spectrum of the system, including initiating communication, facilitating collaboration, defining system requirements, planning work flows, and deploying technology for targeted needs. A systems engineering framework for sustainability has the following elements:

- Focus on the end goal.
- Involve all the stakeholders.
- Define the sustainability issue of interest.
- Break down the problem into manageable work packages.
- Connect the interface points between project requirements and project design.
- Define the work environment to be conducive to the needs of the participants.
- Evaluate the systems structure.
- Justify every major stage of the sustainability project.
- Integrate sustainability into the core functions existing in the organization.

Education is the most powerful tool that can be used toward sustainability. Education personally, economically, or socially is important to improving the

FIGURE 3.1
Sustainability.

bottom line of a business or individual. Sustainability practices save money through awareness and communication while reducing environmental predicaments. Engagement of individuals, whether personally or professionally, will help people to understand their direct responsibility and the effect they have on the environment. When businesses are going through tough times, the number one thing to do is cut costs. Many think cutting jobs is the first answer but instead, they should think first about how to be more sustainable. Innovation mixed with business processes can change the mind-set of people and businesses and reduce incremental costs. Education plays a vital role with these processes because it changes people's ways of thinking.

The methodology is to encourage people to perform certain functions not only while at work but also in their everyday lives. This is a life change, not a flavor of the month. Without making the changes, our entire environment will suffer the consequences.

Education can be the easiest part of sustainability. Education includes engagement, motivation, teaching, and changes of everyday processes, all leading to continual improvements toward our day-to-day lives. Forming teams for sustainability is a great initiative to begin educating others. As a team, each person puts forward his or her own understanding and education of sustainability and participates in a collaborative effort toward the initiative. The beginning process is to form a proper team. The team must be cross-functional, knowing vast areas of the entire environment. A project charter is the next step for this team so there is an executive summary and understanding of what is to be accomplished. Once the project charter is developed, a timeline is imperative so that the focus does not diminish and the priority remains high. Senior management should be given a presentation on the group's objectives and goals along with permission for resources for the effort. A definition for sustainability should be identified next. This definition can vary slightly but should include some of the following terms or ideas as stated in the beginning of the chapter:

> Sustainability is the human preservation of the environment, whether personally, socially, or economically through responsible management of resources, education, and continuous process improvement. Being sustainable corporate citizens will increase sales and profitability by reducing costs and having a competitive advantage.

This statement should be seen as the mission for sustainability. Once sustainability is defined, the action steps are next.

Communication is a must on what sustainability means to the remaining core groups, which then cascades down to everyone involved in the corporation or area. An action plan should come next—how to be sustainable and the steps involved. Each business is different, but commonalities are energy reduction utilizing automated thermostats, turning off computers, utilizing less water, and decreasing gas hot water heaters.

Figure 3.2 shows a world-class manufacturing house with the tools and techniques it should consist of in an organization.

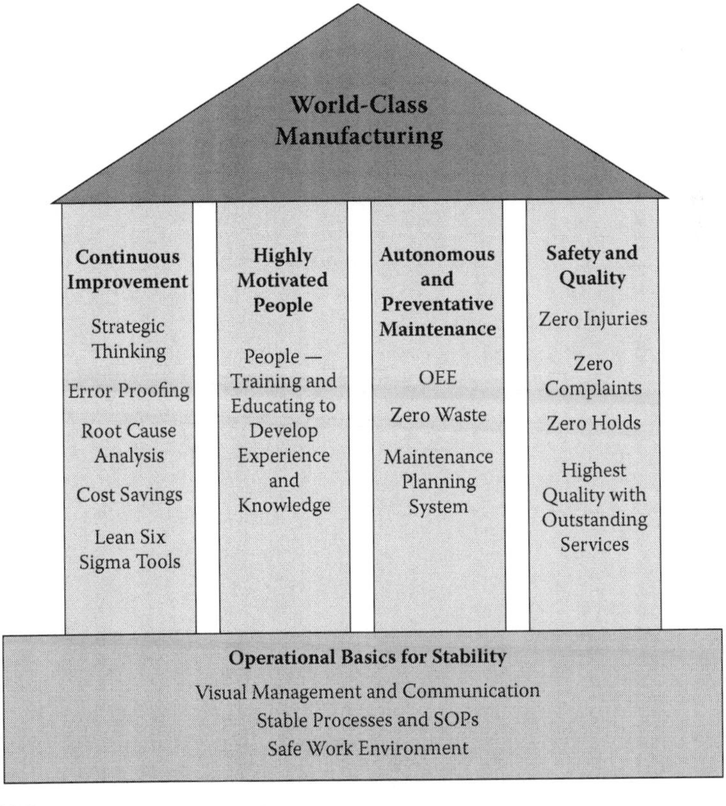

FIGURE 3.2
World-class manufacturing house.

Sustainability Communication

Many technologies are just emerging from research laboratories. There are still apprehensions and controversies regarding their potential impacts.

Implementing new technology projects may generate concerns both within and outside an organization. A frequent concern is the loss of jobs. Sometimes, there may be uncertainties about the impacts of the proposed technology. Proper communication can help educate the audience of the project concerning its merits. Informative communication is especially important in cases where cultural aspects may influence the success of technology transfer. The people who will be affected by the project should be informed early as to the following:

- The need for the sustainability.
- The direct and indirect benefits of sustainability.
- The resources that are available to support the technology.
- The nature, scope, and the expected impact of the technology.
- The expected contributions of individuals involved in the technology.
- The person, group, or organization responsible for the technology.
- The observers, beneficiaries, and proponents of the technology.
- The potential effect of the failure of the project.
- The funding source for the project.

Wide communication is a vital factor in securing support for sustainability. A concerted effort should be made to inform those who should know. Moreover, the communication channel must be kept open throughout the project life cycle. In addition to in-house communication, external sources should also be consulted as appropriate. A sustainability consortium may be established to facilitate communication with external sources. The consortium will link various organizations with respect to specific technology products and objectives. This will facilitate exchange of both technical and managerial ideas.

Sustainability Coordination

Having successfully initiated the communication and cooperation functions, the efforts of the project team must, thereafter, be coordinated. Coordination facilitates the organization and utilization of resources. The development of a responsibility chart can be very helpful at this stage. A responsibility chart is a matrix consisting of columns of individual or functional departments, and rows of required actions. Cells within the matrix are filled with relationship codes that indicate who is responsible for what. The matrix should indicate the following:

- Who is to do what?
- Who is to inform whom of what?
- Whose approval is needed for what?
- Who is responsible for which results?
- What personnel interfaces are involved?
- What support is needed from whom for what functions?

The use of a project management approach is particularly important when sustainability technology is transferred from a developed nation (or organization) to a less developed nation (or organization). In some cases, fully completed technology products cannot be transferred due to the incompatibility of operating conditions and requirements. In some cases, the receiving organization has the means to adapt transferred technology concepts, theories, and ideas to local conditions to generate the desired products. In other cases, the receiving organization has the infrastructure to implement technology procedures and guidelines to obtain the required products at the local level.

To reach the overall goal of sustainability technology transfer, it is essential that the most suitable technology be promptly identified, transferred under the most favorable terms, and implemented at the receiving organization in the most appropriate manner. Project management offers guidelines and models that can be helpful in achieving these goals.

Sustainability in the Workplace

Sustainability in the workplace will increase sales and profitability by reducing costs and providing a competitive advantage. The big questions are how to be sustainable in the workplace. There are three main aspects on how to be sustainable in the workplace:

1. Awareness and education
2. Policies and programs
3. Planning

Some basic questions should be asked first:

- What needs to be done?
- What can be done?

- What will be done?
- Who will do it?
- When will it be done?
- Where will it be done?
- How will it be done?

Four main concepts are sought after when it comes to sustainability:

1. *Economical*—Related to profitability in the economy
2. *Personal*—The effects that are subjected to an individual
3. *Societal*—The informal social gatherings of groups organized by something in common
4. *Environmental*—The setting, surroundings, or conditions in which living objects operate

These four concepts unite together for sustainability because of the parallel relationship they have to one another.

There are some basic cost-reduction strategies that can be used in the workplace. The equations are as follows:

$$\text{Utility cost} = \text{Utility usage} \times \text{Utility price}$$

$$\text{Electricity \$/Year} = \text{kWh/Year} \times \text{\$/kWh}$$

$$\text{Natural gas \$/Year} = \text{Cu. Ft./Year} \times \text{\$/Cu. Ft.}$$

$$\text{Water \$/Year} = \text{Gallons/Year} \times \text{\$/Gallon}$$

Simply put, utility cost reductions are the product of utility usage reductions and utility price reductions. Focusing entirely on usage reductions or entirely on price reductions is not a well-balanced strategy. Figure 3.3 shows that cutting costs is an integral part of making a business profitable.

Understanding the usage of utilities and costs savings that could be incurred is quite simple. In order to correctly estimate utility usage and cost savings, two estimates must be prepared. The two estimates include the utility cost usage prior to the changes and the utility cost usage after the changes.

Annual utility usage savings and cost savings are the difference between the "before" usage/cost estimate and the "after" usage/cost estimate:

$$\text{(Annual ``before-retrofit'' utility usage/Cost)} - \text{(Annual ``after-retrofit'' utility usage/Cost)} = \text{Annual utility usage/Cost savings}$$

Because this is a multifactor analysis (e.g., before and after levels of equipment loading/efficiency/demand, before and after annual hours of operation, before and after utility prices, etc.) there are many opportunities for

FIGURE 3.3
Cutting costs.

errors. Thus, the uncertainty of the utility usage and cost reduction estimates must be taken into account. The conservative approach is to discount the estimates by the associated level of uncertainty.

Cost = Annual utility usage/Cost savings

Utility Cost Reduction Measures can be explained in five simple steps:

1. Efficient operation and effective maintenance of utility-consuming equipment
2. Competitive procurement of utilities
3. Cost-effective expense improvements
4. Cost-effective capital projects and retrofits
5. Efficient design of new buildings and plant expansions

Examples through Simple Means of Sustainability are as follows:

- Computers, radios, stereo systems, and so forth, should be turned off when not in use and unplugged when possible
- Utilize programmable thermostats to save up to 20% annually
- Turn off electricity in rooms not being utilized, for example, think about freezers in factories that are not being used

The resources to be focused on are the following:

- Electricity— No. 1 source of resource consumption
- Natural gas—No. 2 source of resource consumption
- Water/sewer—No. 3 source of resource consumption

Utility costs are one of the largest categories in the annual expense budget for facilities and maintenance organizations. Unlike property taxes and depreciation costs, utility costs can be readily reduced (Figure 3.4). Preventative maintenance is 67% to 75% less expensive than repair upon failure, so it is a measure that reduces net expenses concurrent with implementation. Preventative maintenance need not be applied wholesale; it can be implemented in stages or for selected categories of assets.

There are eight best practices for improved energy efficiency:

1. Increase the efficiency of all motors and motor-driven systems
2. Improve building lighting
3. Upgrade heating, ventilating, and cooling systems
4. Capture the benefits of utility competition
5. Empower your employees to do more
6. Use water-reduction equipment and practices

FIGURE 3.4
Resource management.

FIGURE 3.5
Work as a team to find a sustainable game plan.

7. Explore energy savings through increased use of the Internet
8. Implement comprehensive facility energy and environmental management

So, how do we challenge ourselves and our workforce (Figure 3.5)?

- Educate
- Have a committee
- Come up with a game plan
- Measure successes
- Reward successes
- Continue to improve

Finally, it is known that people feel valued when they are recognized. Therefore, recognition should be used as a tool (Figure 3.6). What does recognition do?

- Creates a "reward" for good energy management
- Helps to engage and motivate people

FIGURE 3.6
Recognition should be utilized for motivation.

- Raises awareness about energy-efficiency opportunities and responsibilities
- Builds support for energy initiatives

It is also important to remember that workplace visual communication is a means for sustainability. The visual communications will enable a workplace to translate information into visible means that are consistent and precise. This visualization will enable strength in the organization, which is undoubtedly present. This will also create a communication methodology that includes the engagement of employees utilizing creative measures to put forth the effort. The belief that sustainment through visual means proves that progress should be used to motivate and encourage others.

Reference

Agustiady, Tina, and Adedeji B. Badiru. 2012. *Sustainability: Utilizing Lean Six Sigma Techniques*. Boca Raton, FL: Taylor & Francis/CRC Press.

4

Empowering Employees

> Power can be taken, but not given. The process of the taking is empowerment in itself.
>
> —Gloria Steinem

Empowering employees is a means to make them and yourself more successful. Needs and wants are interpreted differently when it comes to empowerment. Needing something is important, but once you are able to have it, it no longer motivates you; it is more of an expectation that satisfies. Unsatisfied needs are a means of motivation, which are all interpreted in psychological needs. Psychology plays a huge factor when empowering employees. People want to be understood, valued, and validated, with plenty of positive reinforcement. Once people feel trusted and reassured, they begin to feel empowered and become influential. This is one of the most important assets of communication possible. Motivation comes from wanting to do something because it makes someone feel better. In order to be motivated, one must have the knowledge and skill set for the subject and the desire for success. Being told to do something normally does not empower people because it is not their idea. They want to be listened to and given the choice to make their own decisions. Once what is needed is defined, people can decide if they are empowered to do it because they want to do it. People must agree with the scope at hand in order to be motivated.

Once the scope is agreed upon, commitments on what happens next must be completed in order for the empowerment to continue. For an example, an employee is told that the scope of the assignment is to give safety training to the first-shift employees by the end of the week so that everyone is able to operate the machinery without getting hurt. The employee can be told that if the safety training is completed properly, he or she will be able to be a qualified trainer. The employee now has two motives. One is to train employees for the safety of others and the other is for a possible advancement in his or her career for doing a good job with the training. Explaining what is needed and what the outcome is immediately gives this employee motive for doing the job well. Once the employee has completed his or her end of the deal, the commitment must be kept for the promise that was told. Once commitments are met, employees are able to trust supervisors and managers easily in the future. The first commitment met is the most important commitment because it leads the path to the future. Being proactive to ensure the commitments are met will make the motivation for the employee to succeed in the future an easier task.

Empowering employees also requires the philosophy that employees must be able to trust and respect their leaders. If someone is not respected, it is unlikely that they will be followed. It is a reminder that respect is not given but only earned. Making wise decisions are necessary in order to be respected. A steady path of good decision making with strategic thinking helps motivate employees to trust and respect a leader. That leader then needs to give employees sufficient circumstances for them to make their own decisions. All of the responsibility must not be controlled by only upper management or the leaders. The empowerment comes from letting employees make wise and strategic decisions. Employees do not want to be smothered with what to do next. Instead, ensure that employees are able to make their own decisions without having to check in continuously. Once this philosophy begins, it becomes a habit that becomes human nature. In order to do this as a leader, think about what projects employees are capable of to make decision-making status calls. Once a small project is given with this amount of accountability, in return, the employee is empowering the leader to give more trust with the success of the project. In the long run, this is a major time-saving activity because leaders are not required to be babysitting projects, but instead are empowering employees to make decisions and lead projects. It is important that the communication is effective through the projects and that accountability is organized. Status updates should be presented with questions and solutions. Empower employees to not only bring you problems and questions but they should also have potential solutions. Again, this type of psychological methodology empowers employees to think strategically and make good decisions. This will help them grow as employees and people.

Once one employee sees that another employee is empowered, there will be a domino effect. Other people also want to be empowered and will become motivated to follow the path. As a leader, choosing employees with the skill sets and experience that you trust will help make the empowerment more successful. Choosing highly motivated employees is part of the leadership aspect. It should also be remembered that employees need guidance throughout. When employees ask for this guidance, as a leader, you must give them the guidance they need. It is important to help them think strategically through the guidance so that in the future, the employee is making the strategic decisions from the guidance they have been given in the past. If people have a decision that you would not have thought of, unless there is a major concern, allow the decision to be made and see the outcome. This also empowers employees whether they fail or not to be able to make their own decisions. This methodology will allow employees to grow their own skills and learn from success and failure. Remind employees when they are given a project that it is their project and they own it. Therefore, the decisions they make are their decisions, and you have given them that empowerment. With that responsibility is also accountability that they should be reminded of. It is okay for someone to fail, as long as they have taken responsibility and the decisions that they made do not impact someone else's safety or reputation.

When holding accountability, people must be aware of their goals and the end vision. Giving specific expectations will allow people to perform the job appropriately without straying off task. Holding yourself accountable for your employees is also important. The success of your employees reflects on you as well. You should not blame employees for failures that may result from empowering your employees, instead it should be a team effort. It should be explained how the team came to particular decisions and the methodology of the decisions should be discussed so there is a consistent path to the reason things were done the way they were. Knowing the difference between micromanaging employees and holding them accountable is extremely important. Put yourself in your employee's position when managing them. Think about how you would like to be treated and empowered; most likely the employee you are managing desires the same results. When progress reports are given, asking questions on deadlines and how situations are handled is part of the accountability process. This is also the strategic portion of empowered thinking. Providing feedback also helps to empower employees. Rational feedback consists of both positive and negative reinforcement. This means that there should always be positive aspects reported back as feedback. The negative feedback must be accompanied with ways that the employee could have prevented the problem or different opportunities that could have benefited from the situation. This feedback must not be overly negative, and the negative comments should not outweigh the positives. To be considered valid, the feedback must be given in a timely manner. Specifics should be given with this feedback and only the situation at hand should be discussed. Respect is lost when giving negative feedback on a situation that is not timely or considered in the scope of the project at hand. A great way to empower employees is to ask how they believe they could change the situation in the future. This gives them a chance to think about different ways of being successful versus being told what they have done wrong. Motivating employees to continue to succeed even after a failure is very important. An employee should not want to give up once a failure occurs and as a leader, this empowerment must be provided. It is important to show that the employee has already succeeded by trying in the first place. Empowering employees that the decisions they made may not have been successful but had potential will allow them to want to try again, the next time with success.

The general rule when thinking about empowering employees is to think personally about what empowers you. Remember that empowerment is a psychological aspect that humans need to survive. Employees are empowered by more than just money and time off. People want to succeed and grow personally and that type of empowerment should be sought after.

Empowering employees has the following benefits:

- Improved productivity
- Reduced costs
- Better customer satisfaction
- More strategic thinking
- Best in Class in business

When employees are confident in their own work, they want to be heard. Once employees are confident that they will be listened to, they will speak up and provide their own ideas for increasing productivity to make the business better. With this confidence that their ideas will be listened to and implemented, workers automatically begin being more productive. Once one employee is seen as being productive, most other employees will want to follow, mainly because they do not want to be seen as not pulling their weight.

Once the employees are confident and are working more productively, costs are automatically reduced because more attention is being paid. When people take pride in their work, they do as much as they can to not make mistakes. Just paying attention to details reduces mistakes, which in turn will reduce costs. Mistakes should also be accounted for so there is attention toward the mistakes. Once people are held accountable, again mistakes begin to be reduced because they do not want to be known as the ones who always have problems. It should be known that human nature allows for mistakes to happen. To reduce some human types of mistakes, mistake-proofing techniques such as Poka Yokes should be used (see Chapter 11, "Continuous Improvement Toolkit").

Customers are naturally satisfied when they get what they want. Putting focus on projects increases customer satisfaction. When customers receive the product they want in the manner they expected, they are happy. The more mistakes or defects customers find in the products they receive, the unhappier they are, and the more likely they are to go to a competitor for their needs. Customer satisfaction occurs when customers know there are improvement projects toward what they are purchasing. When a customer is asked for feedback and knows that their opinion matters, they automatically feel empowered that their decisions matter. Customers become more loyal when asked for this type of feedback. Customers also appreciate consistent service. This consistency is the same thing that employees need. Employees are internal customers and should be treated the same as external customers. The next process in the system is always passed to the next "customer," so all personnel in the workforce are customers. Gaining feedback from employees is the same as gaining feedback from external customers. They feel empowered that they were asked for their thoughts and that their feedback could make a difference.

Once employees see that their opinion is making a difference and changing the workplace, they will begin to think of more ideas to help the business.

This incorporates strategic thinking because they are always thinking of ways to make things better. The employee begins to not always do things the way they always have, but instead start challenging the status quo and being strategic. Empowering employees to want to be strategic thinkers will gain different and innovative ideas.

All of these items will make the business Best in Class. If the business is the best at what it does, it is considered Best in Class. This philosophy is extremely important because customers make judgment calls very quickly, especially when they are unsatisfied. If customers are not satisfied, they have options and will use those options. Customers normally have a multitude of options for where they receive their goods or services. Therefore, being Best in Class in the business will ensure customers will go to the Best in Class business because it is known to have quality products that are consistent. Complacency must be driven out of the business in order to grow. Once the business is Best in Class in what they do, they must innovate because competitors will catch on and try to be Best in Class themselves. To be competitive, there must always be Continuous Improvement on the current ideas at hand. Even if something is working well, it should be sought after to find something better that customers would prefer and would not mind spending additional money for. Customers will pay for quality and innovation. The breakeven point on what they are willing to pay must be considered when being innovative.

Empowering employees does not just mean giving people extra money or "things" to make them happy. They want a great environment to work in, which in turn, will help make the business a realistic place to work that is successful and beneficial. Employees want to enjoy what they are doing, want to be cared about, and want to have freedom to make their own decisions. Employees have been known to pick a workplace that is more fun and open-minded over a place that is micromanaged where employees are not listened to even if the latter pays more money. Employees want to go home at the end of the day feeling like they made a difference. Once this need is satisfied, an employee feels empowered. The ability to be flexible is a priority that is just as important to an employee as a salary. They want to be able to change the status quo and reduce the complacency of their everyday life. Once employees enjoy what they do, this enjoyment rubs off on others. Enthusiasm is contagious. If one employee is enthusiastic about their job, they will convince others to feel the same. The same goes for negativity in the workplace, unfortunately. If there are only negative attitudes, other people will also pick up on the negativity. This principle is especially true from managers. The top-down approach is important for empowering employees. This philosophy entails that top managers manage their employees by giving them a sense of empowerment and satisfaction. Each of the managers should do this to their own managers and eventually all employees will be affected. This approach will then cause the bottom-up approach, where the employees being managed are empowering managers to change their ways by seeing that their employees are happy when given flexibility, empowerment, and decision-making skills.

Micromanagement in this case is the worst approach to take because it is seen as negative. People feel when they are told to do something in particular they are less entitled to want to do it versus when it is their own idea. They also do not want to be asked for status updates every hour. They want to be trusted for their work and want to be able to tell their managers when they have completed a milestone. When employees are happy with the work they have done, they want to let their managers know what they have done. Once they do this, the praise portion is extremely important in order for the empowerment portion to be complete. Praise helps employees feel good about what they have done, which makes them want to do good things over and over. This philosophy reassures employees, making them confident in their abilities. The way employees are treated is essential to the success of the business. This treatment must be consistent across the board for all employees and should have the top-down and bottom-up approaches. When all individuals feel they are being trusted and are being treated well, they are essentially empowered no matter what position they are in.

Genuine openness, courtesy, listening, and respect ensures success in a business by empowering employees to want to make a difference. This management by respect is much easier to use than management by fear. This ability to utilize proper people skills increases the ability for changes for the positive to occur. Encouragement and enthusiasm make a workplace fun. All successful leaders have one thing in common, which is their people skills. People skills is what it takes to empower employees to work for you and work happily and successfully. The diagram in Figure 4.1 shows the philosophy for successfully empowering employees.

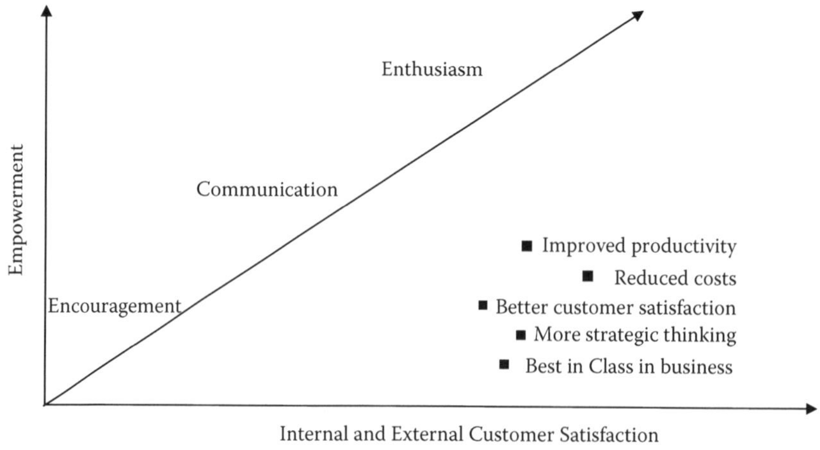

FIGURE 4.1
Empowerment for benefits to internal and external satisfaction.

Another tool to ensure employees are engaged is to ask them for their feedback. This can be done through an anonymous form, as shown in Figure 4.2. The importance of the Project Recommendation sheet is to have a full round of feedback. Employees want to communicate about the recommendations they submitted and whether their idea was implemented and the reasoning behind it.

Empowerment is a tricky tool and should be managed properly. An empowering manager should possess the following traits:

- The demonstration that people are valued
- A vision is shared between leadership and all other employees

Continuous Improvement Project Recommendation

Employee Name(s): Date:

Position: Shift:

Department:

Project Department:

Project Location:

Type of Project (Circle):

Process/Job Improvement *Safety* *Waste Management* *Plant Organization*

How do YOU believe this will benefit Company XYZ?

Employee(s) Name: Approving Supervisor Name:

Employee(s) Signature: Approving Supervisor Signature:

FIGURE 4.2
Project recommendation sheet.

- The goals given to the manager or the business are shared and transparent
- Trust is ensured
- Decision-making skills are taught and people are empowered to make decisions
- Project management skills are up to par
- Feedback is given frequently
- Problems are solved and not just told
- Listening is demonstrated
- Employees are rewarded and empowered

Most employees feel empowered through more than just money and it is important to seek what each individual employee desires and treat them with the respect they deserve. Once employees are empowered, they will continue to do a job well and will want to go to the next level to ensure success. Empowered employees will help managers succeed by being successful themselves.

Reference

Covey, Stephen R. 2004. *The 7 Habits of Highly Effective People*. New York: Free Press.

5

Project Managing Employees Including Your Boss

> There are managers so preoccupied with their e-mail messages that they never look up from their screens to see what's happening in the nondigital world.
>
> —Mihaly Csikszentmihalyi

Positivity in your workforce is imperative in order to maintain sustainable relationships, especially with upper management. Managing your boss in a way that is not overpowering and shows effective communication will better your relationship in a positive way. This is done through listening, comparing perspectives, and your understanding of one another. Once you learn about your employees and your manager, it will be more easily understood how to adapt to the different environments and personalities. By letting employees and management know you are learning and listening for the better and changing your actions, you will begin gaining the respect that you deserve. Ensure that you are giving positive feedback, but that it is not overpowering. If you have relationships that need work, do not rely on others to make the first move; you must initiate improvement steps. Always treat anyone you deal with by respecting them and always being courteous whether you feel they deserve it or not. Do not engage in negative conversations with others regarding employees, upper management, or your boss; having negative and behind-the-scenes conversations can only lead to trouble. You also never know who will move up the ladder in your corporation, meaning your employees may be your boss one day.

To be a successful project manager, it is important to be project managing the appropriate projects. Developing a business case that is solid for the projects is important. Many projects begin without having a good rationale or business case. Documenting the business case and then getting buy-in from senior management is the way to begin to start a good project. The project must fit the business vision and goals. Everyone must agree that the project will impact the business positively in the way the business wants to grow. Senior managers should be able to identify key priorities along with key stakeholders for the project. Once the resources are allocated, the risks should be assessed before beginning the project. It is important to know reasoning behind failures of the project as a whole or particular parts to the project. If there is a budget in the project that may not be approved, these types of risks should be identified early as well. The risk analysis at this point in time should be performed at a high level until greater details are established. Having a high-level return on investment (ROI) also normally helps gain buy in for the project. Even though the ROI is high level at this point in time, it will help justify the project to move forward and help with the solid business case.

Once a business case is identified for the project, the project must be defined with a project charter. The project charter should be as specific as possible using SMART (specific, measurable, attainable, relevant, and time-bound goals) (Figure 5.1).

Responsibilities of accountable personnel and their involvement must be identified at this point in time. The accountable personnel must also confirm their resource needs for the project. The project manager must have buy in from senior management for this.

When the project begins, it is important to have a kickoff meeting so that each person is on the same page about what is to occur. The sponsor of the project must be the designated person who leads the kickoff meeting. This kickoff meeting should have an agenda, the key personnel involved, and a discussion of how often the team and the stakeholders for the project should meet. During key phase gates, another meeting is required. An estimation of key milestones should be delivered to establish a baseline for the project timeline and completion date. If critical paths or milestones need to be made and require resources, they must be identified up front as well. The communication of the project will ensure the success of the project. All personnel should know exactly what is happening for the project whether good or bad so they can do their part to follow up or add value toward the project.

The project manager is required to know all of the key aspects that are happening for the project along with any hurdles which are involved. The project manager should not implement any of the milestones, but instead should serve as an agent helping the project steer along successfully while

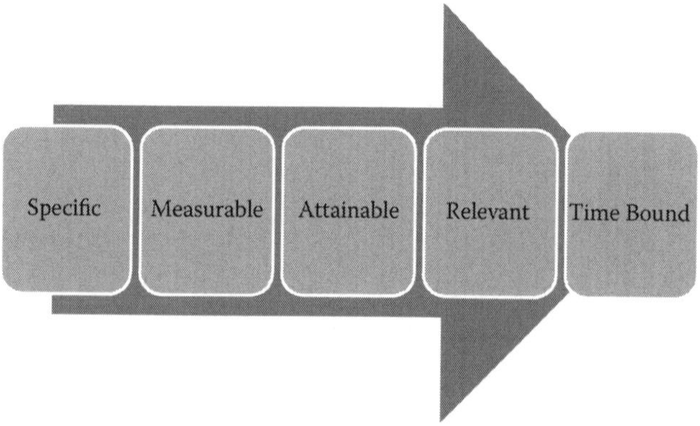

FIGURE 5.1
SMART goals.

communicating key data to the group and senior managers. The project manager must document all this information and be prepared. Having an agenda and documents for each meeting is essential to an effective meeting. The meeting should include a description of the key milestones and when they will be completed. The milestones that are not completed and what could be done to alleviate issues within the project should also be known. Barriers will be told to the project manager, who is to support and provide resources to eliminate barriers and ensure that everyone's top priorities are being completed. It is also important that the top priorities of all resources are aligned. Regular meetings should occur and be organized by the project manager. An agenda should always be sent out ahead of time and the work behind the agenda should be sent to the project manager prior to the meeting. Meetings should be about updates and progress reports versus work in meetings. The work should be performed behind the scenes and the updates of the work should be presented in the progress update meetings. Always refer back to the project charter to ensure that the project is in scope of what was planned. Communicating the milestones and the project status is important to being an effective project manager. It is also important not to embellish the truth. If the project is not going well, the reasoning behind it should be discussed and known versus waiting until the closure of the project. The key stakeholders should be able to help with understanding the barriers and closing gaps.

Once the project finally comes to a close, it is important to again have a wrap up meeting or a project closure meeting. All key stakeholders and team members should be present at this meeting. The message for the meeting should include the following:

- The anticipated scope
- The actual progress and goals of the project
- The final successes of the project
- The key learnings of the project
- What can be done better for the next project

Ensuring that the scope was met and was not derived upon is important for the project manager's success. Understanding the risks involved and associated with the project is also part of the end goals for the project manager. Not all items will be completed at the finale of the project. It is important to document the items that remain and what control plans are needed to maintain the success of the project. The final communication and celebration of the project must occur to motivate the employees who took part in the project. The project closure of finalization must be performed and the accountable people must be noted. A successful project will have a documented project report that shows all the goals, successes, milestones, key learnings, and finalization of the project.

The Project Management Process

If done correctly, the project management process will ensure that your employees are appropriately managed and your boss is not only kept up to date on your progress but project managed as well.

To summarize, the following steps to the project management process should be completed for successful projects to be implemented and will be described thoroughly:

- Brainstorm a project that is aligned with the business vision and goals.
- Ensure there is alignment within the business, the stakeholders, the key team members, and yourself for what you want achieved.
- A clear vision statement is created.
- All employees are held accountable for their appropriate job tasks.
- The project is in scope, and milestones are met and timed out appropriately.
- All information is documented and progress is tracked accordingly.
- Communication takes place for current progress, tasks to be completed, and information for any people involved or who have a need to know about the project.
- Timeliness is accounted for by all members and for all milestones.
- The final project is completed, sustained, evaluated for learnings for future practices, and successes are celebrated.

During brainstorming for the project, the matrix in Figure 5.2 or one that is similar should be utilized.

The project management process has four main vital processes (see Figure 5.3).

The brainstorming process is to think about challenges within, the impact of daily activities, associated timelines, and priorities.

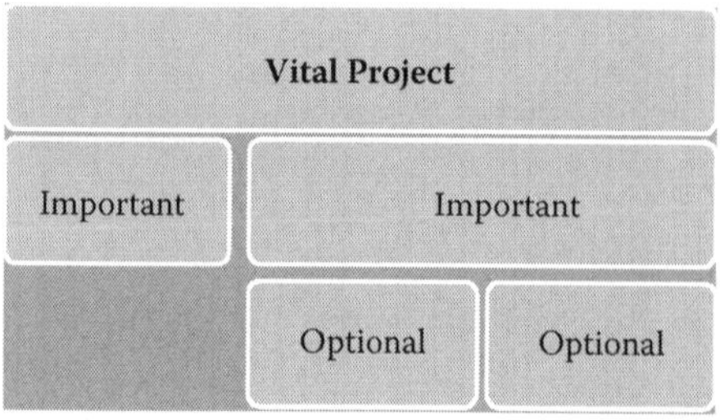

FIGURE 5.2
Brainstorming projects.

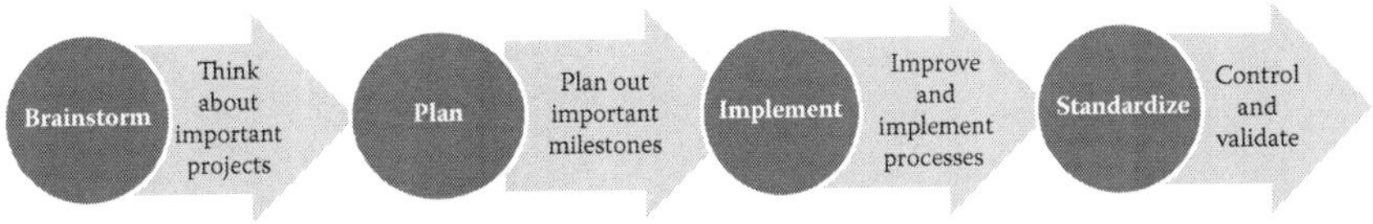

FIGURE 5.3
Four main project management processes.

The strategy of the project must be thought about after the brainstorming. For success, the vision and goals must be looked upon and alignment must be present. Strategic projects follow the process shown in Figure 5.4.

When understanding alignment, a Voice of the Customer questionnaire should be applied. This process includes asking key stakeholders and people involved what matters to them.

Voice of the Customer

- What is the business vision?
- What are the business goals?
- Why is this project important to you?
- How will this project impact you?
- What can be done to ensure the success of the project?

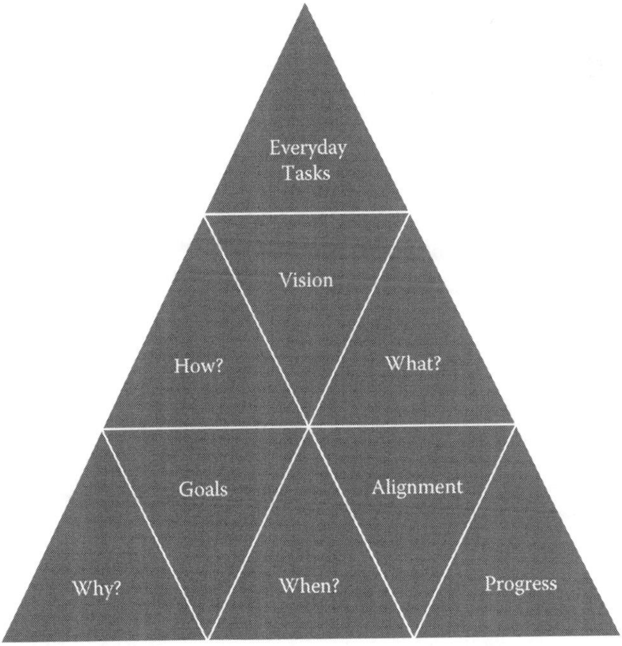

FIGURE 5.4
Strategic project pyramid.

- What timelines matter most to you?
- Do you have any information that will help with the success of the project?
- How can the sustainment of the project be ensured?

The vision statement and planning is the next process in the project. The vision statement should involve the following:

- What is the purpose?
- Describe the project.
- What are the anticipated goals?

A mission statement defines the organization's purpose and primary objectives. Its prime function is internal to business goals. It will define the key measure or measures of the organization's success. The prime audience is the senior management and stakeholders.

Vision statements also define the organization's purpose, in terms of the organization's values rather than bottom-line measures. Values are guiding beliefs about how things should be done. The vision statement communicates both the purpose and values of the organization. For employees, it provides a direction on how they are expected to behave and inspires them to want to work hard and do a good job. Shared with customers, it shapes customers' understanding of why they should work with the organization.

Accountability should occur in the project. An easy way to ensure accountability is being kept is by documenting tasks, the people accountable, the date, and the progress. A simple spreadsheet for accountability is shown in Figure 5.5.

The planning process is essential for the implementation of the project. It is important to know where to begin, to plan out scheduling and milestones, and to identify key resources. It is also important not to involve all participants if they are not needed for all meetings. Some participants will just be informative participants unless needed at a particular time for their subject matter expertise. The planning process has several steps that include:

- Prioritization
- Brainstorming with exploration and the discussion of potential successes

Area	Priority	Department	Project	Action	Follow Up	Completion	Savings	Responsible	Due Date	Cross-Functional Team	Notes
Labor	1					0					
Waste	1					25					
Labor	1					0					

FIGURE 5.5
Accountability spreadsheet.

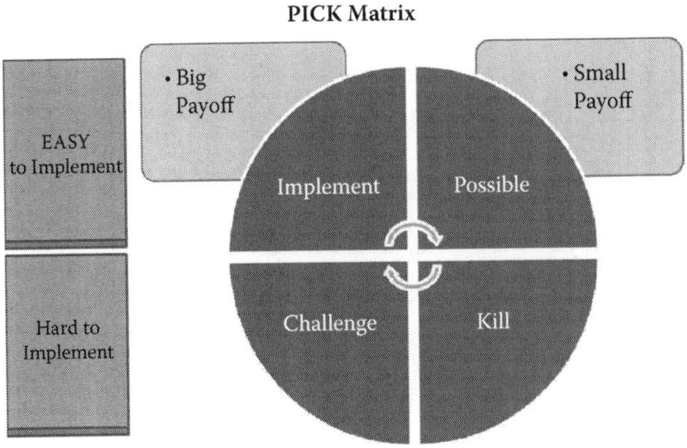

FIGURE 5.6
PICK matrix.

- Projects put into in scope categories, mini projects, or tasks
- Major tasks should be discussed for milestone evaluation
- Timelines should be determined
- Critical steps should be discussed for milestone achievement and barriers
- Resources and budgetary constraints should be discussed

A potential brainstorming scheme will include a PICK (possible, implement, challenge, kill) matrix, as shown in Figure 5.6. A PICK matrix is where projects can be put into categories to understand the payoff value and the likelihood of success for the project. Any project in the implement category should be taken care of immediately and the possible projects should be discussed for potential resource constraints. The challenge projects should be tried to be made more attainable to be moved into the implement categories. Any projects that are in the kill category should be revisited to find a spinoff project.

The prioritization, brainstorming, and mini projects can be put into the PICK matrix and discussed during milestone meetings. The constraints, importance, and planning should be discussed during this time when determining mini projects.

The remaining steps in the planning process should be put into a project task map shown in Figure 5.7.

During the planning phase, it should be known what decisions should be acted upon and which ones should not be in scope for the project. To ensure success, in scope ideas should meet priorities, have low risk, realistic timelines, and appropriate resources, and be in alignment with key stakeholder's goals.

Project Task Map

Project Title ____________ Start Date ____________ Project Completion Date ____________ Actual Completion Date ____________

Project Description ____________

Success Criteria ____________

	Priority	Project Task	Responsible	Start Date	Finish Date	Actual Finish Date	Planning

FIGURE 5.7
Project task map.

The prioritization of the performance is an important factor during this planning phase. The brainstorming should be thorough but include key stakeholders. Manageability during the planning phase will ensure the project will actually get completed well and on time. Accountability of key resources must be agreed upon during the planning phase. PICK matrices and project task maps are resources that help with the planning and prioritization process.

The implementation is the next process in project managing. It is important to know how to manage and track workloads of yourself and of others while communicating progress with the team. Barriers will come up during the implementation phase and should be overcome. Delegation is a key part for project managing employees. The implementation will include planning for the future. The planning should be daily and monthly including effective meetings.

Effective meetings consist of the etiquette shown in Figure 5.8. Focus must be met for meetings to be effective (Figure 5.9). An agenda must be sent out before the meeting and must be followed during the meeting (Figure 5.10). Time management must be ensured during the meeting (Figure 5.11). Being cognizant of other people's time is important and courteous. An action plan should be included during the meeting (Figure 5.12). Refer to Figure 5.5 that includes an accountability spreadsheet. Decisions must be made during meetings (Figure 5.13; also see Chapter 8). Being courteous is not just polite but helps meetings move along precisely (Figure 5.14). In conclusion, meeting etiquette is a key practice for project management (Figure 5.15). A meeting agenda template is shown in Figure 5.16, which should be followed for proper meeting successes and courtesy during meetings.

The implement phase is effective if the planning phase was properly integrated. Revisions to the plan will occur, but should be minor. Scheduling

Effective Meeting Etiquette
Send out the agenda in advance.
Make sure the correct individuals are in the meeting.
Start and finish on time (50 minute meetings).
ACTION PLAN
No meeting inside a meeting.
Stay on task.
Formalize the issues and have clear follow-ups.

FIGURE 5.8
Meeting etiquette.

appropriate meetings and having proper agendas is important for the implementation piece. Delegation must occur during this phase and people must be held accountable for their tasks. Review meetings will ensure the team is on track and communication must be completed for all stakeholders and team members. Effective follow up will ensure the project stays on track.

As a project manager, interruptions will occur. In order to bring the group back to a focused session, the project manager must regain the attention back to the group. Ideas that were not in scope must be put in a "parking lot" but still documented. Guidelines must be given to the group so intentional interruptions do not occur frequently. These guidelines may include having a no phone, computer, technology rule or ensuring that no one has any conflicts during the appropriate meeting time. Scheduling the meeting in advance and with the proper resources with a timeline and agenda will help the meeting to move along properly. It is also courteous when sent an invite to respond with a commitment. If committing to the meeting, prework must be done. Attachments and action items should be sent to the meeting attendees during this time. This way people come to the meeting prepared. If you are unable to attend a meeting, it is courteous to respond with a reason why, but also ensure that your work is handed into the project manager. Another meeting should not need to occur if an employee cannot make the meeting. The work should be given to the project manager beforehand. A meeting status notice should be sent to all the invitees after the meeting to ensure that everyone understood the action items and progress.

FIGURE 5.9
Focus for meeting etiquette.

One Point Lesson

Subject	**Meeting Effectiveness**

Done by: Meeting Effectiveness Team

Type	[X] Basic Knowledge	[ ] Improvement Case	[ ] Problem Case	[ ] Activities Transfer

Applicable to a new employee? (X) yes ( ) No

Have and Follow the Agenda

A-G-E-N-D-A

FIGURE 5.10
Agenda for meeting etiquette.

FIGURE 5.11
Time management for meeting etiquette.

FIGURE 5.12
Action plan for meeting etiquette.

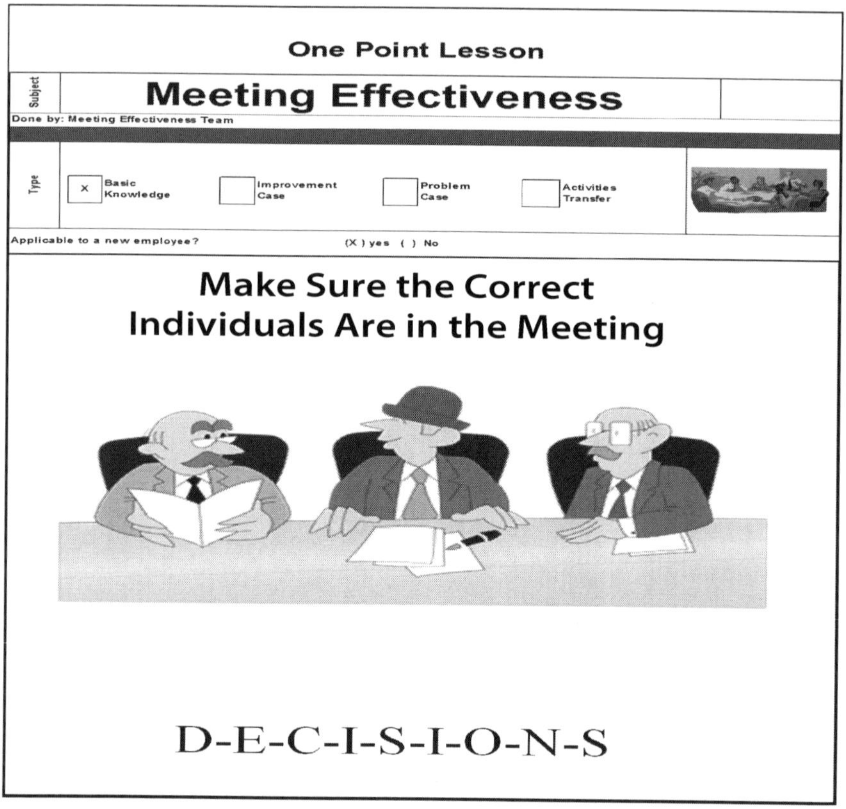

FIGURE 5.13
Decisions for meeting etiquette.

One Point Lesson

Subject: **Meeting Effectiveness**

Done by: Meeting Effectiveness Team

Type: [X] Basic Knowledge [ ] Improvement Case [ ] Problem Case [ ] Activities Transfer

Applicable to a new employee? (X) yes ( ) No

No Meeting Inside a Meeting

C-O-U-R-T-E-S-Y

FIGURE 5.14
Courtesy for meeting etiquette.

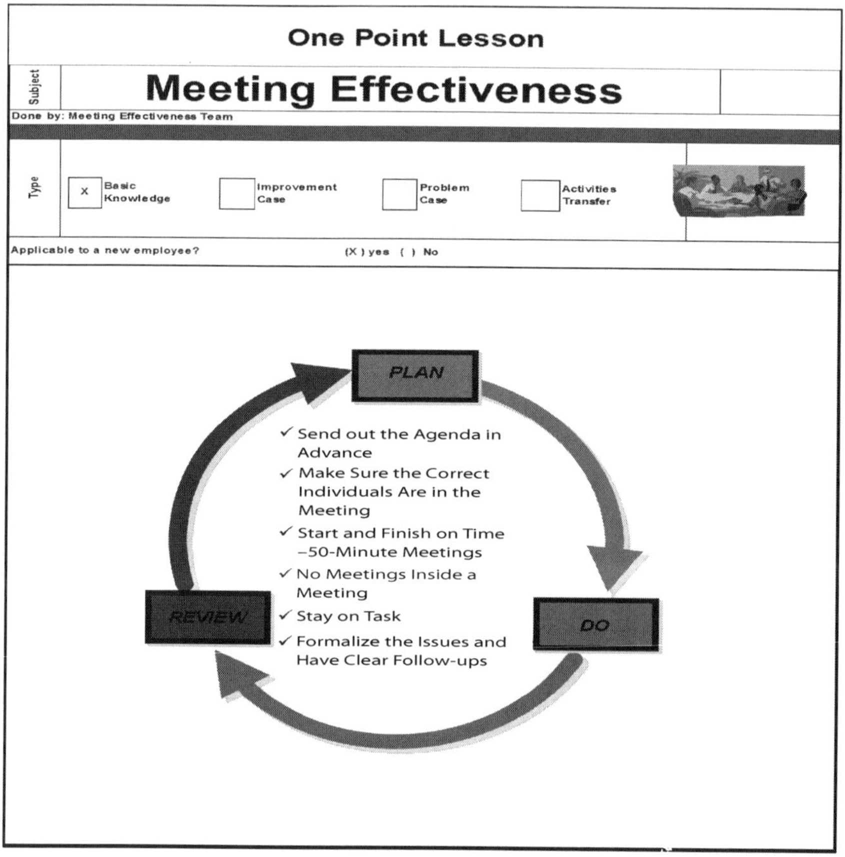
One Point Lesson
Subject
Meeting Effectiveness
Done by: Meeting Effectiveness Team
Type
X Basic Knowledge
Improvement Case
Problem Case
Activities Transfer
Applicable to a new employee?
(X) yes ( ) No
PLAN
DO
REVIEW
✓ Send out the Agenda in Advance
✓ Make Sure the Correct Individuals Are in the Meeting
✓ Start and Finish on Time –50-Minute Meetings
✓ No Meetings Inside a Meeting
✓ Stay on Task
✓ Formalize the Issues and Have Clear Follow-ups

FIGURE 5.15
Cycle for meeting etiquette.

MEETING AGENDA

DATE/TIME:
LOCATION:
CONFERENCE BRIDGE:

[INSERT MEETING NAME] – MEETING AGENDA *(please come prepared to discuss agenda points)*	*TIME*
• "EHS" Environment, Health, and Safety	2 MINS.
• Topic/Objective/Consensus	
• Topic/Objective/Consensus	
• Topic/Objective/Consensus	
• Topic/Objective/Consensus	
• Recap and Open Items (moves to next meeting)	
TOTAL	

[INSERT MEETING MATERIALS]

MEETING NOTES/ACTION ITEMS

[INSERT MEETING NAME] – MEETING NOTES/ACTION ITEMS	*DATE*	*OWNER*
• Action Item		
• Action Item		
• Action Item		
• Action Item		
• Action Item		
• Action Item		

Checklist **:**

- ***Agenda Sent Out***
- ***Correct Individuals Invited***
- ***Start and Finish on Time***
- ***No Side Meetings***
- ***Stay on Task***
- ***Formalize Issues and Have Clear Follow-Up***

FIGURE 5.16
Meeting agenda template.

Project Evaluation

Project Name:

Project Manager:

Date:

Was the project statement met?

Did the project stay in scope?

Was the plan for project resources adequate?

Was the project schedule realistic?

Was the project budget met?

Were key milestones met?

Was there alignment with key stakeholders and team members?

Were the appropriate people involved in the brainstorming phase?

Did the planning go well?

Was the implementation of the project on target with the plans?

Are there plans for sustainment of the project?

Does the project meet the business vision?

FIGURE 5.17
Project evaluation sheet.

Standardization of the project must occur for the entire project to be successful. The project is not truly completed without validation of the project and controlling the outcomes. During the standardization of the project, the following questions should be asked:

- Was the project successful?
- What can be improved for future projects?
- Is the documentation complete?
- Were the project successes aligned with the business vision?

The project evaluation can be completed by ensuring that the budget was met, the documentation was complete, and the stakeholders are satisfied with the results of the project. A project evaluation sheet should be given to team members and stakeholders so that improvements can be made in the future (Figure 5.17). A project evaluation sheet can include ratings if needed.

The finalization of the project is completed once the project has action items toward the validation and standardization of the project and the team has celebrated the success of the project. A documentation of the project must be complete for validity and to help in the future when finding projects in scope.

Reference

Mind Tools. Mission Statements and Vision Statements: Unleashing Purpose. http://www.mindtools.com/pages/article/newLDR_90.htm.

6

How to Get People to Trust You

> Learning to trust is one of life's most difficult tasks.
>
> —Isaac Watts

Trust, like respect, is not given but earned. Trust takes time, patience, and consistency. Even if you are trusted a majority of the time, the one time that you do something untrustworthy, your trust will be severely compromised. As a leader, people listen and watch closely. Speaking about what you want to do and actually doing what you said you were going to do, go hand in hand. Being ethical in all situations automatically gains trust, because you are doing the right thing. It is important to always do the right thing even when nobody is watching. Once this becomes a habit, people will see that being ethical is a norm and not just a trait you want people to perceive in you. Trusting yourself also gains trust from others because the positive aura is passed along.

What traits does it take for someone to trust you?

- *Loyalty*—Through good and bad, your consistency is measured and you can always be counted on.
- *Honesty*—Sometimes even when unpleasant, the truth is given when needed. There are no lies within.
- *Accepting*—Judgments are never to be used when dealing with people.
- *Supportive*—Support through endeavors without judging or showing off your own success.
- *Helpful*—Taking time to help someone when needed.
- *Understanding*—Understanding your situation even when you are not in it. Being able to know what someone is going through whether they speak up or not.
- *Listening*—Being able to listen and not being preoccupied shows you care.
- *Respect*—Treat someone as you want to be treated no matter his or her job title or position.
- *Commitment*—Keep your promises.
- *Generosity*—Showing you are generous out of your own goodwill that you have nothing to benefit from.

- *Positive attitude*—If you believe you can, others will believe you can.
- *Relationships with others*—When others see the relationships you have in a positive sense, they will trust that you are able to have that relationship with them as well.
- *Responsibility*—Making smart decisions and backing up your own actions will show that you are a responsible person.
- *Self-discipline*—People want to see that you have the courage to make good decisions and are proactive without being told what to do.
- *Vision*—Have a positive vision and share that vision.

The qualities that make up trust make a good leader that individuals want to be represented by. People become followers of good leaders through their good qualities because they want to be like the leaders they trust. Think about the attraction of some leaders you respect personally. Do they have the aforementioned qualities? Recognizing what trust is, how to be trustworthy, and knowing what personal characteristics to change is what it takes to be a good leader. It is important to have a good leader who you respect and that you find the qualities in that leader that make them trustworthy. With the traits you find, you can begin incorporating those qualities personally to make yourself the leader that other people respect. Remember, trust takes time, so finding the good qualities will not come overnight, but instead needs to be practiced over and over so that they eventually become habits.

Leaders are looked upon almost all of the time. Therefore, any situation that leaders are put in are watched and criticized. It is important to deal with circumstances to the best of your ability. What does this mean? Sometimes situations will occur that are not pleasant and you may react in crisis mode. Showing instead that you can take a step back, analyze the situation, and make a decision based on your knowledge will show that even during bad times, the proper actions were taken. Showing character during hard times makes a person more respected because they are able to overcome situations that most people would not know how to handle. Believing in yourself during those difficult times will make others believe that you will make the right decisions no matter how hard the scenario. Trusted leaders do not fall apart during both success and failure; they are calm and composed throughout. This strong character makes someone look positive and respected. People will begin to trust you more during difficult situations, because they trust you know good versus bad and will stay away from the bad with a uniform decision.

Commitment is such an important term when it comes to trust, because people want to see that you are personally committed before committing yourself to others. Think about the commitments you make personally and how they affect your life. This can be as simple as promising yourself that you will take that morning jog every day before work because you have committed to it. Telling your peers about your commitment will make them hold you accountable as well. Once they see your commitment through actual actions, they see that you have personally committed. Once people see that you have committed to a personal goal, they want to make their own commitments or trust that when you make a commitment you are able to follow through. Commitment is tested by actual action and not words; it is important to follow through once you have spoken about your commitments. Not following through on commitments automatically makes the trust factor go down in a relationship of any kind. Commitments are not only about showing that you do what you say but also shows achievement. You commit to goals because you want to change and be more successful. Commitments show that you believe in yourself and have powerful goals that you are determined to accomplish. Commitments are measured upon as well. If you commit to jogging every day like you said but only end up jogging 5 of the 7 days, even though the benefit is still virtuous, the commitment is broken. Commitments need to be realistic and not just performed to show off or gain respect from others. Therefore, think about the commitments you make before you make them. Think about your everyday life and what you want to achieve. Will you benefit from working out 5 times a week versus 7? What will the extra 2 days do for you? Is it more realistic to start with the goal of 5 days and then work up to 7 when you feel more comfortable? It is those decisions that gain respect. People respect people who can make a commitment that can be maintained. Remembering that commitments are about personal pride as well as trust will eliminate the uncertainty of others that you are committing just to make others want to trust you. Measuring commitments will ensure that you are on track, and telling people about commitments will help people hold you accountable. The success factor is much higher when you measure your successes and ask people to hold you accountable because you do not want to let yourself or others down. Doing your best for you and mostly you will also ensure that your commitments are maintained. Remembering that you are making commitments to be a better person, not to show off, is extremely important and can be seen by others vividly.

Communication is key during commitments because you are being effective by showing your goals whether it is for the business or yourself. The goals can be clearly shown by your successes. Communicating commitments is easy, to be effective it takes the following:

- Explain your goals and the reasoning behind your goals
- Be truthful with your goals
- Be realistic with your goals but also show your long-term goals or wants
- Confide in people
- Seek input from others to ensure that you are gaining respect

Generosity is another aspect that people respect. People want to know that you care about other people and not just yourself. Generosity should be out of gratification for oneself and not done for praise. Ensuring that you have other people's interests in mind shows your generosity. If you can put

FIGURE 6.1
Listen.

someone else first, it shows that you have their best interests at heart and shows that you are trying to accomplish a value.

Remembering that trust is a two-way street is very important to gain trust. People not only want to listen to you, but they want to be listened to as well (Figure 6.1). Listening helps people learn and gain wisdom. Understanding what other people are about and not just what you care about makes you a broader individual who understands several scopes. Listening creates clarity in different situations and can help make difficult decisions easier by listening to the past experiences of others. Listening involves really understanding what someone is saying and why. It is important to know why the person feels the way they do or is doing what they are going to do. Listening is an act that does not judge, but takes the information in and puts it in a level of empathy. Listening involves being focused on an individual and being on their level. People want to have a common ground, and listening shows that you have put aside your thoughts and are focused on them. Listening and communicating are important because people want to know that you listened to what they had to say. By ensuring that you are properly taking in their message, questions about what the person is telling you helps show that you are involved in the conversation. People also want feedback. Feedback should be given positively whether it is what someone wants to hear or not. It is important not to be harsh when giving feedback; tell the reasoning behind your feedback. If someone is hurt by your feedback, they in turn will not come to you again and will not trust you or future conversations with you. Emotions play a large role when listening because people are emotional when they communicate. Ensuring these emotions are thought about and understood when listening helps people trust you. Having a sound basis behind your feedback will show that you have prior experience and are giving advice based on positive outcomes from the listening you have done in the past.

A positive attitude is seen by all and is a choice that is made. When someone tells you something, you immediately have two options:

1. Be negative
2. Be positive

Your attitude needs to reflect positivity in order to gain trust in others. The past may have made you negative because bad things have happened, but instead, take the past situations and put a positive spin on them. Not allowing your attitude to deteriorate due to negative instances will make you a more positive person in general. People perceive positivity easier than negativity and find it encouraging. People automatically are drawn to positive people because they give out positive energy and vibes. They make people feel good. Positivity is about taking in information and finding the best scenario, having goals that make sense, and communicating the positive outcomes that are happening.

Positive thinking comes from three main aspects (Figure 6.2):

1. Removal of negativity
2. The surrounding of positivity
3. Identifying problems and establishing goals and a vision

Negativity is like positivity, with too much influence, it becomes a behavior that is instilled upon. Removing yourself from negativity or responding to negative behavior with positive behavior will have benefits. Experience from the past will show that there are positive outcomes to even negative scenarios. Remembering this concept to remove negativity will provide the choice to remove frustration or unneeded worry and doubt. Utilizing other people's positive behavior or independently seeking out positivity will help remove negative vibes or negative behavior.

Surrounding yourself with positivity is as easy as the law of attraction. Positivity attracts positivity. Being proactive ensures that positivity is gained with rewards attached. Surrounding yourself with positive thoughts and people ensures that your focus and thoughts are automatically positive and will become a habit. The negativity will eventually not even be a thought

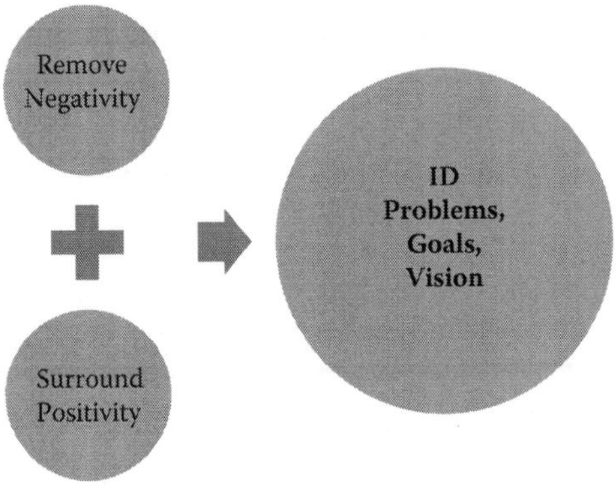

FIGURE 6.2
Remove negativity + Surround positivity = ID problems, goals, vision.

because it is not thought about anymore. It is important not to let negativity control you but instead to allow optimistic thinking to take over.

Knowing how to get along with people is important because it ensures that this behavior is passed along to others. Everybody is different, so it is important to know how to get along with different individuals no matter the personality. This instills trust because people see that you are adaptive to others and diverse people and behaviors. This shows a caring mentality and people want to know they are cared about. All people have a few things in common:

- People want to be cared about.
- People want to feel special.
- People want encouragement.
- People desire a vision.
- People want to know what is in it for them.
- People have emotional needs and need positivity.
- People want to be successful.

These are all truths of people and in order to be trusted, people want to be accommodated to these truths of themselves. Treating people as the individuals they are and not a group will make your interaction with them much more successful because they can see themselves as a single person you are dealing with. Different people have different personalities, so treating people the same may only have a 25% success rate. Conforming to different personalities will change this success rate to 100%. Knowing and loving people as individuals will show empathy in others and show that you have an ability to find the best in people. Helping people is part of the empathy process keeping people interested in what you have to do or say. The foundation of relationships is based on this trust and respect people have for you. To improve your relationship with others, improve your relationship with yourself by having a positive outlook and mind. Strengthening bonds with others will help yourself mentally by giving you a positive heart and mind.

Being a responsible person has benefits to being trusted as well. Success is about responsibility and good leaders are responsible people. Letting people count on you and instilling these values in others will make other people be responsible as well. Responsible people get the job done almost all of the time, which is why they are in a position of such stature. They have repeatedly been counted on with successful results and the trust they have maintained from the past makes them the responsible individual they are now. Understanding what needs to be done and focusing on the goal will help the success of making the goal a reality. Responsible people do not just get the job completed, but they go above and beyond what needs to be done. They are proactive and have an agenda in mind for success. Success and excellence drives responsible people and gives off the positive vibes discussed upon earlier. Responsible people also continue to do the things they find right no matter what situation they are put in. Responsible people are sometimes given harder situations to deal with because they have previously responded positively and that behavior is desired again in the upcoming problems. Responsible people stay in it for the long haul and do not take subpar as good enough. They strive for excellence and have a positive mentality that leads to them being trusted.

Responsible people have their own self-discipline. They know what needs to be done and when it needs to be done. They do not need people telling them what to do all of the time because they are proactive in finding the right things to do at the right times. Self-discipline is about wanting what is best because it gives off positivity and excitement. People want to be successful out of good nature. Self-discipline is completed by the following attitudes:

- Developing a vision
- Following priorities and making commitments
- Being disciplined
- Removing negative behaviors and putting positivity in place of them
- Completing the task at hand to its fullest ability while going above and beyond

Having a vision is one of the most important things to have in life. It gives meaning to what needs to happen and why it needs to happen. Leaders have a passion and courageous behavior that must have a vision in order for the follow through to take place. Knowing one's own vision is the beginning aspect before adhering to other people's vision. If you know what you want, it is easy to make a vision for other aspects of your life because it must fall in line with your own personal vision. A vision starts within oneself based on past experiences and the need for success. A vision shows accomplishment and has milestones along the way. Success is measured by ensuring that the vision was maintained and responded to in the most effective fashion possible. A vision shows others the responsibility that you have taken upon yourself. This vision is normally transferred to others due to the positive feedback that makes people trust your vision and the vision you have shared for the group. Understanding the vision and ensuring that you are being truthful within will gain respect and leadership qualities from others and within oneself.

After breaking someone's trust and confidence in you, it will take a lot of patience and determination for the trust to be regained. With perseverance and determination, it is possible to turn around a person's disappointment in you and make your relationship better than before.

Failed trust will cause future trust to be ruined. Regaining trust involves setting the bar much higher than it was at the beginning. If trust was once lost, the person should not be punished for the mistrust; he should just be informed of the doubt behind trusting again. Earning trust again must happen once a failed situation has occurred. Trust is also found when discussing other people's trust. If someone sees that you are telling other people's secrets, they may also not trust telling you their secret because you may do the same thing behind their back. Trust should be kept when given the promise. Trust involves one party relying on another; it relies on the other party not harming or failing the trustee. The person essentially has confidence in the person they trust when they are giving them the belief that the information they share will stay between them. There are three steps to regaining trust once it has failed:

1. Convince the party that you are to be trusted again by showing them with your actions. Ensuring that the circumstance does not happen again is very important. Understand what was done and do what you would do if you were in the other party's shoes. Take patience to understand that pain is encumbered during mistrust. Sacrifice knowing that you are not to be immediately trusted.
2. Apologizing for being untrustworthy must happen. Be specific about the apology and even document it if needed. Do not put pressure on the other person to forgive you on your time.
3. Move forward after the apology occurs. Give space and time after a trust issue has occurred. Remember to forgive yourself and give the other party time to deal with the mistrust. Accept the fact that the trust may not be rebuilt as well as it was in the beginning.

Gaining trust is easier at the beginning when mistrust has not occurred. According to *Forbes*, there are nine ways to gaining trust in a hurry:

1. *Make friends in high places*—Having references behind your trust shows you are to be trusted and many already trust you.
2. *Don't be a dummy*—Ensure you have the right information, but do not pretend to have information that you actually do not have just to sound like you are always informed.
3. *Admit your mistakes*—If you want to seem truthful, fess up to your mistakes. Liars normally do not apologize.
4. *Be confident*—Be self-assured in yourself without having doubt. Do not make up lies to appear better.
5. *Don't be tense*—A trusted person is calm and composed.
6. *Make eye contact*—A person finds value in you being able to look at them in the eye and speak truthfully to them.
7. *Learn the language*—If there is a language barrier, you do not have to learn the other language but make attempts to communicate so they are able to understand you and see you are trying to communicate with them.

8. *Pass out compliments*—Make people feel good about themselves by telling them positive things that will make them smile.
9. *Be predictable*—Ensure you do the same things that people trust, such as making meetings on time, calling when you say you will, following through on commitments, and so forth.

Trust is earned over time, so being reliable and showing that you speak out of honesty goes a long way. When you are trusted, you are considered dependable, considerate, loyal, and friendly, which are all aspects people want. Trust is an important part of all relationships and businesses, and should not be taken lightly.

Reference

Forbes, http://www.forbes.com/2010/07/22/trust-relationships-confidence-opinions- trust.html.

7

Changing the Status Quo

Status quos are made to be broken.

—Ray Davis

Change involves changing yourself along with others. Change can always be for the better, but no change at all can never lead to better situations. Change management relies on the understanding of why things are done and why people are comfortable. Managing others through change processes is needed to change the status quo (Figure 7.1). This type of change needs guidance, encouragement, empowerment, and support.

What is the status quo anyway? The status quo is defined as the existing state of affairs. In Latin, the meaning is "the state in which." So maintaining the status quo means to keep things the way they currently are. Some people have the mentality of "If it isn't broken, why fix it?" This methodology can never lead to change or success because even if things are going well at the present time, in due time other changes in the world will come into effect that make things not go as well as planned. In a business setting, every corporation strives to be the best it can be in what it does. The competitors then try to beat the Best in Class corporation by doing things differently and better than their competitor. Eventually, the Best in Class corporation is the one that produces the most satisfying changes, but the change must be present in order to satisfy their customers. Keeping things the same way rarely satisfies customers because customers become complacent and bored. Humans desire change and innovation. The desire to be different motivates others to change the status quo. Being complacent normally means being safe and avoiding controversy. Even though this is a safe measure, it will not end with a Best in Class way of doing things because complacency becomes tiresome and the thrill of excitement is taken away.

Effectively implementing change involves frequently incorporating new competencies. New competencies will help all employees engage in further education and training. Once they are used to this mentality, changes become good things and they are always anticipated. The changes will begin from simple items to make work easier to strategic thinking where daunting tasks are eliminated in means of more technologically advanced methodologies.

People are afraid of change because it makes comfortable ways uncomfortable while changing the normal way of operating. What should be taught about change is the fact that new skill sets are being taught, which opens

FIGURE 7.1
Status quo.

doors and increases continuous education. The first reaction of asking someone to change can be taken negatively because they are being told to stop doing what they have always been doing. People can take offense thinking that what they are doing is not sufficient enough and they become quite territorial about their work. To help people understand why the change is needed, the approach methodology should be given along with the reason the change is desired. Most of the time change is needed to do things more efficiently while speeding up tasks that need to be completed. Understanding this change is imperative in order to embrace the change.

Spotting resistance to change is important because negative connotations to change must be eliminated. Spotting change comes from listening to gossip or what people have to say about why things will not work. Observations of people will tell whether the employees are resisting change as well. If employees are being extremely negative, arguing with others no matter what statement is made, missing deadlines or meetings, or not attending or reading up on change assignments, normally means they are being resistant to change. This resistance must be minimized by addressing the situation. The resistance can only be minimized if the employees have trust in the changes. If the employees are involved in the decision making, they become less resistant to the change because they had a say in the changes made. The clear-cut communication also gives visibility on the changes happening with less ambiguity on why the decisions were made. Building positive relationships helps minimize this change by having positive bonds with people working together. The multiple positive attitudes end up being passed on to others with negative attitudes making the entire workforce change. The open exchange of ideas must be able to occur in order to change. If employees are afraid to bring up their ideas, they will continue to resist change because they are not being heard. It is essential to gain employees' trust and allow them to speak their minds. Once they describe the feelings they are having and the reasoning behind them, it is important to understand the position they are in. Minimizing their fears they have will slowly gain their trust and ensure their fears are not going to come true.

Knowing what is driving the change and how it will affect individuals and the organization is an important step to acceptance of change. Once the change has been adapted, people will begin to like the change and be willing to make more changes for the better.

Being a role model during change eases the difficulty of the change (Figure 7.2). People want to be facilitated through change transitions, and need guidance and coaching. To be a role model, it is important to put yourself into the other person's shoes that needs to complete the change. Thinking about all the reasons the change can be bad will help with the understanding of how others are feeling and coping. Giving the business vision on why the change needs to happen will also help guide the scope of the project. Never accept "That's the way it has always been done" as an answer. This answer is an excuse and not a solution.

The combination of skill sets will help with challenging the status quo. Different people have different ideas and combining the ideas to come to a uniform decision will help the group to find strategic opportunities. The group needs to be flexible, and one individual cannot overpower the conversation or ideas being given. Adaptability of other people's viewpoints, schedules, concerns, and styles should be taken into consideration. No one idea is right. Showing other people as a leader that you are open to change will

FIGURE 7.2
Change.

FIGURE 7.3
Timing of change.

help other people appreciate your values and anticipate the change. Limiting expectations for a perfect workday scenario should be considered in order to accommodate other people and give them the time and attention they need during the time of change. People need a great deal of reassurance during changing times, and as a leader that reassurance and commitment need to be present daily. Leading with values such that others will respect and follow you is important because people do not follow others just because they have the title of manager. The vision needs to be not just told, but shared and agreed on by all. Leaders need to show that they are not changing the status quo for selfish reasons but for the benefit of the project and the business.

Timing of change is important as well (Figure 7.3). People need time to understand the change, and let the new thoughts and ideas sink in. During this time, many questions will incur, and employees should be able to speak with their managers about the questions they have with an open mind-set mentality. Employees need to be a priority to managers. Without employees, managers are not managers. Competing priorities will lead managers to have little time for their employees. Clarifying expectations of changes and what you as a manager are working on will help others understand what they are dealing with. Employees should never feel as though they cannot talk to their managers. Once this fear is instilled in their minds, the trust has already been broken and mending the trust is harder to fix than starting with an open mind-set mentality. If priorities do change, the transparency should be given promptly with explanations for the changes in priorities. All people are emotional, so giving facts about situations rather than reacting toward emotions is important.

Anticipating successful results to maintain a positive outlook contributes to optimistic vibes given to others, which can be contagious. The approachability of the changes can be taken from different aspects:

- Positive Thoughts Approach
- Humoristic Approach
- Realistic Approach
- Open-Minded Approach

Maintaining a positive attitude is the most important aspect of changing. Positivity will gain respect and will allow others to see the brighter side of the picture. Humor can ease tensions and bring people together by seeing the lighter side of a situation. Humor helps during the beginning of relationships being established to show there is a fun and real side of perspectives. Being realistic helps people not to gain false expectations of what could occur and ensure that honesty is well presented. Being open-minded helps people see all versions of what could happen and appreciate different points of view.

Resistance to the Status Quo

All approaches and reactions are different and should be seen with an open mind. A great deal of people resist the status quo versus embracing it. This is a problem that should be overcome and should be seen with an end in sight. In order for people to begin embracing the status quo, the following should occur. Ensuring that people understand the benefits of the change helps to fight resistance. Incorporating different views and ensuring that all perspectives are taken into consideration will help decisions to be made and accepted. Motivating and encouraging others that changing the status quo for the better will help resistance as well. Listening and understanding through the resistance will also show that the different perspectives are appreciated versus having a "my way because I said so" attitude. Patience is extremely important through resistance. It may take people more than an hour or a day to embrace change. Showing that you have patience through the change will help ease the situations and show that there is a realistic approach behind the methodology. It is important to show, share, and ensure others believe the vision being handed at large. Alignment is key to changing systems and resistance. Creating lists of pros and cons can also show the vision and reasoning behind changing. During the list creations, the current and new ideas should also be listed to show why things were performed in the past. Finally, positive reinforcement is a must through resistance of change. Not all people have natural self-confidence. Giving people the confidence they need is easy and is a coaching practice that makes good leaders. Publicizing wins helps people's self-confidence grow and shows that working through difficulties can be managed appropriately. Flexibility through this management must be learned by employees, managers, and leaders.

Utilizing Known Leaders to Challenge the Status Quo

It is known who self-leaders are by their ability to influence others and engage in activities. It is to a manager's benefit to utilize these leaders to help the change. Assessing the current structure of who the natural leaders are, whom the resistance comes from, and who the followers may be will make the process easier to understand and manage. Simple questions can be asked:

- Who do people listen to?
- Who will adapt to change easily?
- Who will resist change?
- Who will help with the alignment of change?
- Who will motivate others?

Once the support structure is understood, the utilization of the proper people in the proper aspects will align the process change management. Identification of trouble areas will help leaders to be realistic in what items may not go as planned. Utilizing the known leaders to hand out information needed will also solve problems. These people will assist in supporting the structure, the processes, and maintaining the vision. Goal alignment with known leaders must be established before the leaders are utilized to motivate others. Once these goals are shared and agreed upon, these leaders will help change management drastically.

Communicating Change

Communication is the most important aspect to successful changes. Communication prepares others for what is to happen, creates shared and agreed upon visions, and builds relationships within groups. Communicating change should consist of the following:

- Communicate specifics of what changes are occurring and why.
- Explain the importance of the change to the business and to yourself individually.
- Explain the urgency of the change to occur and the associated timeline.
- Communicate the downfalls of what could happen if the change does not occur.
- Publicize wins of work well done or employees who have embraced change.
- Encourage and motivate others on why to change.

- Prepare information and updates on results, changes, and progress.
- Communicate in person.
- Admit not knowing information and provide a means to show that you will find out the information.
- Be truthful even about the downfalls.
- Make expectations clear.
- Be transparent.
- Do not show favoritism, be consistent.
- Identify roles and responsibilities.
- Learn from mistakes and be transparent about why mistakes occurred and how to prevent them in the future.
- Capture Best Practices and learning techniques that should be learned.
- Recap the last communication topics and outcomes.
- Obtain employee and management support.
- Increase awareness of new initiatives.
- Utilize risk management techniques such as Cause and Effect Matrices, Root Cause Analyses, and Failure Mode and Effect Analysis (see Chapter 11, "Continuous Improvement Toolkit").

Successful transformations of workplace change occur by asking basic questions and appropriately combining them for the proper answers:

- Who?
- What?
- Where?
- Why?
- When?
- How?

How to change and what to change are the big factors that explain the success of change. Once the establishment of these measures are in place, the implementation process is simple. Successful workplace changes occur with the proper foundation and framework that has been established through management and employees through a shared vision. People must drive these visions together with their different skill sets. The cross-functional groups working together as teams is what drives successful change. Without that piece, the rest are just words and pictures. The real change comes from the people combined together.

It should be remembered that information sharing is a powerful means of communication. Information given to one another helps build teams by sharing known information and making strategic decisions from the data. A company's culture is based on this information sharing. Culture is based on beliefs and values, and the combination and agreement of the two. Wanting to increase productivity and efficiency is part of the culture. If all people believe in this, the vision is the same. If some people are not bought in, the culture shift needs to take place. Once the culture has the same enterprises of beliefs, the overall vision can be achieved by all entities working on the same goals due to their beliefs.

According to The Steve Roesler Group, there are five clear-cut messages that show it is time for a change. These five qualities must be looked for and are defined next:

1. *People whom you trust strongly believe you should make a change*—If multiple people who you are close to think it is time for things to be different, it may be a time to listen to them. They may understand you and see the changes may actually benefit you.
2. *You're holding on to something and just can't let go*—If a particular situation is continually on your mind and you simply cannot let it go, this is seen as a signal. If this is bothersome for you, instead of mentally abusing yourself, a change could ease your mind.
3. *You feel envious of what other people have achieved*—Jealousy is an evil beast, but it may be able to help you better yourself. If you are envious of someone or something, instead of being envious, you are able to change your ways to become more successful instead of simply being jealous. Taking action toward bettering yourself reduces jealousy and makes you proud of yourself.
4. *You deny any problem and are angry in the process*—Anger is seen as a symptom of denial. Looking for help from someone or being able to help someone in need reduces the anger. Increasing communication will help the problems be mitigated by being able to admit there is a problem at hand. Having an open mind to the problems will encourage you to make a change for the better.
5. *If you do absolutely nothing, the problem will continue*—Without addressing the situation or holding people accountable, no change will be made. This is because it is not known that the problem is bothering you, or nobody wants to address the problem at hand. Addressing the situation must happen for change to occur, but must be in a professional and calm manner, or will be seen as negative. Being honest and up front will help others see what changes can be made and will help yourself be able to change for the better with the communication received.

Tackling workplace status quo uses these simple methodologies:

- *Communicating where wasted time is being spent*—This involves giving facts on costs and data behind efficiencies. Visual management of key performance indicators (KPIs) will address where targets and goals are. See Chapter 11, "Continuous Improvement Toolkit" and Chapter 9, "Visual Management."

- *Address costs from doing work repetitively*—Discuss extra labor hours and extra materials spent due to mistakes and having to correct these mistakes. See "Mistake Proofing" or "Poka Yokes" in Chapter 11, "Continuous Improvement Toolkit."
- *Communicate the need for efficiency improvements*—All companies and people need to be more efficient. Communicating the need to increase efficiency because capacity restraints are present is an easy way to show the goals of efficiency improvement are needed. Showing competitor's efficiency numbers or predictions will help communicate the baseline of where the efficiency should be.
- *Show that change is easier than people think*—Nobody likes to have a "flavor of the month" of how to do or address something. People must see change as easy and positive. It is important to show change is innovative and teaches others creative measures. Continuous education is imperative to growing and changing the status quo will invigorate these behaviors by having an intelligent workforce that is up to date on technology and teachings.
- *Show that change is a habit*—Change should be a daily habit and not a way of doing something for a particular time. Incorporate change into people's everyday lives so that change becomes something easy and normal. Show changes are good for personal and professional growth, and must be incorporated in their work life and personal life.

Dealing with change is simple if people trust why changes are occurring. Explaining that change is needed in order to be competitive and expand while being innovative must occur for the change to take place. Management's beliefs and values need to be shared as a combined vision versus a top-down approach so that each individual is on the same page about change.

There are four main changes that occur in the workplace:

- New products or services
- Organizational changes
- New management
- New technology

Explaining the need for the different changes and the rationale behind the decisions made will help buy in for the change. Sometimes the change does not seem to be for the better (for example, layoffs). It is important when communicating change, even change that is not desired, that a positive outlook can be seen from the change. The best way of communicating any type of change is through factual information to the best information given. The majority of the time, the changes taken place are positive and affect the following:

- Cost
- Processes
- Culture

Cost changes should be discussed as being for the better in order to be competitive in the marketplace. If cost changes do not occur, the market demand will go toward the cheaper good especially if there is a good choice of quality. Process changes must be made to have Continuous Improvement. Increasing efficiencies will increase the market for being competitive as well as accomplishing an easier way of doing things. Innovation should be sought after to keep up on the latest technologies. Cultural changes are normally the most difficult because it changes the way normal operating procedures take place. Explain that cultural changes are needed so there is a combined vision between all people in the workforce and there is no divide between management and other employees.

Implementing successful change involves utilizing the proper resources whether those resources are things or people. The plan then needs to be established and agreed upon so there is reasoning behind the changes being made. The changes then need to be implemented. Many times people think of great improvement ideas but do not fully implement them, making the sustainment of them impossible. Finally, communication is the end of the change management. All steps through the process must be communicated to all people and must be agreed upon in order to be successful. Effective communication as stated in Chapter 1 involves communicating often, giving reasons behind changes; explaining who was involved in the changes; and giving status updates of what changes have occurred, what will occur, and the effect on the business from the changes.

References

Gebelein, Susan H., Brian Davis, Kristie J. Nelson-Neuhaus, Carl J. Skube, David G. Lee, Lisa A. Stevens, and Lowell W. Hellervik. 2004. *Successful Manager's Handbook.* Roswell, GA: PreVisor.

Green, Robert. 2012. Tackle the Status Quo in Your Workplace. Cadalyst. http://www.cadalyst.com/management/tackle-status-quo-your-workplace-15262.

Managing Change. Reference for Business. http://www.referenceforbusiness.com/management/Log-Mar/Managing-Change.html.

Roesler, Steve. 2008. Earn Your "Change Chips" Early. All Things Workplace. http://www.allthingsworkplace.com/change_and_transition/.

8

Decision Making

> Make a decision already!
>
> —Tina Agustiady

Making no decision is much worse than making the wrong decision. Decision making is challenging, but shows initiative and drive.

Every morning you wake up and make a decision. This decision can be simple like pressing the snooze button or deciding not to take that morning jog you were planning. Whatever decision you make, whether simple or difficult, will affect your day and possibly even your life. Think about the pros and cons, extreme or minor:

Situation: Deciding against a morning jog

Pros: Sleep an additional 30 minutes, minimize risk of injury, won't be as hungry

Cons: Most likely won't take jog later, will still eat just as much as with the jog, may become habit leading to never working out, considerable health problems in the future, won't fit into summer bathing suit

This simple decision already had several pros and cons. Some were short-term issues and others will have a long-term effect. This same decision-making problem occurs daily at work. Procrastination is normally the main key of decision making. Decisions are made in the near future to put an item off, but the long-term effects are not thought about in detail. Decisions made influence everything that follows after, and creates either problems or opportunities. Decisions should be made to improve communication and help influence others to be positive decision makers as well.

The first step of decision making is determining the end goal or vision. When deciding in the morning whether to take a jog, think about what it is you are aiming for. Are you preparing for a major run and this is your practice? Are you trying to get in shape for a big event? Are you trying to fit in a certain outfit? Are you trying to lose a certain amount of weight? Are you trying to be healthier in general? That small decision of taking that one morning jog will affect all the outcomes of your goals and visions for the future.

The next step is understanding if the right decision is made and the impact it has on other people. Think about if you had a running partner instead.

If you had a running partner, would this decision to jog in the morning be simpler? How would this decision impact someone else? Would you end up convincing the partner to skip most of the jogs? Would you feel accountable to hold up your end of the bargain and ensure you kept your commitment of jogging? Would your partner hold you accountable? These are the goals that need to be clear when making your decisions.

Decision makers need to keep the following in mind:

- Short-term goals
- Long-term goals
- Impact on the organization
- Impact on other people
- Conflict resolution
- Communication with others impacted

Consulting with others when making a decision may help by understanding other people's expectations. You will easily be able to understand whether people find your decision acceptable. Also, accountability is kept when involving others in your decision-making process. It should be kept in mind, however, that it will be difficult and cumbersome to ask others for help with every decision you need to make. Some decisions need to be weighed out individually and a sound reasoning behind the decision should back up the end outcome. Figure 8.1 demonstrates that success versus failure are ultimately two different avenues. Risk analyses can be performed when making decisions. The three areas of risk that need to be considered are:

- Severity
- Occurrence
- Detection

This type of risk analysis can be performed utilizing simple tools such as "Failure Mode and Effect Analysis," as described in Chapter 11, "Continuous Improvement Toolkit."

To ensure buy in, some type of rationale must be made toward decisions. This rationale must include being aligned with business priorities

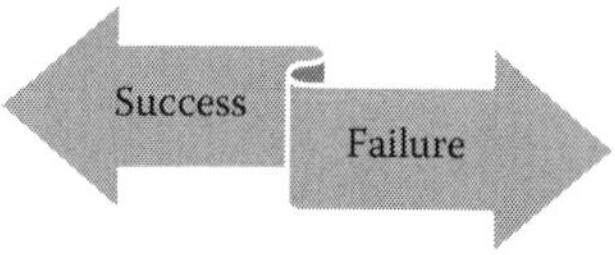

FIGURE 8.1
Success versus failure.

TABLE 8.1

Decision-Making Criteria

Decision	Pros	Cons	Costs	Risks
A				
A				
A				
B				
B				
B				
C				
C				
C				

and being aligned with all associated pros, cons, and risks. An example of decision-making criteria is shown in Table 8.1.

When making decisions, data must be used. Data-driven decisions are the best decisions because they are decisions made on facts versus opinions. Buy in comes much easier when facts are involved.

Consider the following scenario: An industrial engineer is trying to remove extra labor that is included in the production line when the machinery fails to do its job properly. The operations manager does not want to remove the labor because there is a potential that the machine will fail. The industrial engineer wants to remove the labor because it is costing the company $100,000 in the operating budget.

When the industrial engineer does the analysis, he or she should gather facts. The following facts and data can help make a sound decision:

What percentage of the time does the machine fail?

Can an error-proofing technique (see "Poka Yokes" in Chapter 11, "Continuous Improvement Toolkit") benefit from ensuring that the machine does not fail?

What costs are being incurred by the machine failing?

Are the costs more than just the labor? Rework, extra materials being given away, customer dissatisfaction?

What are the business goals?

Will the employees agree to the changes?

What are the pros and cons?

Where is this in the priority list? (Operating budget cuts, reduction of labor, etc.)

Having this data will lead the operations team to make the decision quickly and efficiently with the facts they have at hand. Documenting these facts and data and being prepared when asking the decision will make the process easier and without conflict. This is also part of challenging the status quo of how things have always been done.

Once the information is gathered, the information needs to be analyzed. This can be performed in an individual or group setting.

Identifying the issue at hand and the end goal is the most important aspect of the decision-making process.

It should also be reviewed what needs to happen for a proper decision to be made. This analysis will help understand if further investigation needs to be performed or possible resources or machinery is needed. All the unknowns need to be mentioned as well so that decisions are made on the facts at hand and the risks from not having particular information are evaluated. If assumptions are made, the assumptions should be documented. Possible questions will be the outcome of analysis and assumption. The questions that could be asked should be evaluated in this analysis phase so that all bases are covered. If all of the information is not at hand for the decision, a decision could be postponed. It is important to gather as much information as possible before requesting a decision. When there are questions or ambiguity, the problem at hand should be addressed versus the question posed. The questions should be addressed as soon as possible, but should not prevent a decision unless the data is absolutely required to make the decision. The risk behind making the decision without information should be addressed so delays are minimized.

Sometimes subject matter experts are needed to make proper decisions. Ensuring the data is captured from the proper knowledge and expertise will also back up decisions or analyses performed. More than one subject matter expert will make data look more presentable and well approached. This will also show a logical approach to how the analysis was performed. Thinking outside the box with new innovative ideas can be presented during this analysis, but a rational option or backup plan should also be presented so that the data is logical and valid. The innovative ideas should be determined as an option that needs to occur if the business goals and priorities are aligned with making innovative decisions and resources can be given. Before presenting the final analysis, the data should be evaluated again to ensure it is rational and logic. The data should also be presented as just data and not opinions trying to sway personnel to choose one idea or another. The decision makers will make the decisions based on facts and the compelling business need versus opinions on what should be done to make a better case. Assumptions should be made into hard and soft assumptions. These assumptions are considered hard and soft for the following reasons:

Hard assumptions—Based on fact or previous circumstances

Soft assumptions—Based on opinions and not validated by more than one person, may not be a reliable source

When making decisions, a team approach is widely used. This approach is used so that there is justification on decisions. It shows that multiple decision makers were in agreement versus one power player.

Decisions made on the fly are normally the cause of issues that occur in situations. This type of decision making should be diminished so that decisions are made with proper reasoning. Going back to the example of taking the morning jog, it can be easy to make a quick decision in the morning not to take the jog because your state of mind is altered. You may be fatigued in the morning or feel lazy. The factors that have already been evaluated should be looked at when this natural problem occurs. The pros and cons need to be addressed as quickly as possible. Also, if it is known that excuses will be made, identify items to do to ensure that the excuse is not made such as having a running partner or putting out clothes for the jog. Putting an alarm clock away from the bed could help get you out of bed. These simple decisions we make daily are also the basis for the complicated decisions we make in the business world.

Once a decision is made, responsibility and accountability must be taken toward the decision whether good or bad. Because of this, the thought process behind the decision-making process must be thoroughly evaluated. Concerns should be addressed and other people's views and understanding must be thought about. The end goal must be thought about when taking responsibility of decisions made. The impact of other people could be tremendous. Past circumstances should be evaluated when making decisions. These past circumstances should not deter people from challenging the status quo or changing the way things have also been done, but the risks associated with the decisions from the past should be weighed. Discussing what went wrong in the past and why will help evaluate past circumstances. Once it is defined what went wrong in the past, evaluating what could be done differently so that problems do not arise will help mitigate similar issues from the past.

Timely decisions are also imperative. Making a decision versus no decision is looked upon in a very critical way. Quick decision making can be looked upon as a hasty decision. Long decision making is looked upon as procrastination. Timely decision making is looked upon as seeking urgency, utilizing appropriate information, and evaluating other people's values.

Sometimes hasty decisions are made because there are deadlines that need to be met. It needs to be understood that decisions should not be procrastinated, but making decisions without valid input and output data will affect the goal and vision. Discussing inputs and outputs toward the process can also help with brainstorming proper decisions to make. While looking at the inputs and outputs with the processes, the end customers and beginning suppliers should be evaluated as well. This type of analysis is called a SIPOC evaluation (Figure 8.2) and is explained in Chapter 11, "Continuous Improvement Toolkit."

Suppliers	Inputs	Input Specification	Process	Gap	Outputs	Customers
Raw Bone	Thread	1.0–1.1	Materials come from warehouse to XLine		Raw Material	Wally
Pet Doo	Material	2.0–2.9	Material fabricated by machine X		Fabricated Material	Tget
Cryer	Bone	4.1–4.9	Toy sent to machine Y		1/2 processed toy	Kroman
Whiner	Cushion	.8–1.0	Materials put together by machine Y		Toy complete for inspection	Giant Store
Happy Pup	Squeaker	5.5–6.5	QC Process	**Manual Process**	Toy complete for inspection	Pet Peeps
	Rope	10.1–10.9	Rework		Toy complete for inspection	
	Treat	2.4–2.10	Decorate product		1/2 processed toy	
			Package		Fully processed toy	
			Ship		Toy ready for customer	

FIGURE 8.2
SIPOC.

When a decision must be made due to immediate attention or urgency, the risk analysis should be referred to. All the information needed will not be possible to be sought after in this situation, but what information is absolutely imperative needs to be evaluated. Finding out what additional information can be found to make a more sound decision will help evaluate the risk of the situation. The timing of finding additional information is most likely critical for the decision, and should be questioned whether the decision would be different with the additional information. Anticipation of the outcome in this case should be considered when all the information cannot be gathered in a timely manner. Again, the pros, cons, and risks of each situation should be evaluated so that the decision is made valid after the information needed is found.

When delays are made due to information not being available, it should also be evaluated; what type of information was missing and how much time did it take to gather the information. Involving people who may have a negative connotation toward the decision-making process will help the communication process. Before involving these personnel, the big picture should be looked at, to see who needs to be involved. People affected by the decision along with key business personnel are the people who should be involved in the key decision-making processes. It is important to not only involve upper management in these types of decisions but also the personnel in the business the decisions affect. The effect of the personnel in the business should be evaluated to discuss obstacles or measures to gain buy in for success.

When a decision is finally made, the communication of the decision is critical to gaining buy in. The communication should occur in a timely manner and the reasoning behind the decision needs to be told. It is important for anticipated questions to be thought about and a retaliation mind-set not to be used when there are negative comments regarding the decisions. The communication should come firsthand to all personnel involved so rumors, which may lead to potential problems, are not started. Regular updates regarding the decision will also make the decision made an easy transition. The clarity in the decision-making process and who was involved needs to be conveyed. It should also be explained why the personnel involved in the decision-making process were chosen and why.

The following process should be completed when making decisions:

1. Defining the problem
2. Gathering data
3. Evaluating data
4. Deciding who to involve in the decision-making process
5. Involving appropriate personnel
6. Discussing goals, priorities, and business needs
7. Defining associated pros, cons, and risks
8. Discussing people affected by the decision made and the consequences behind the decision
9. Defining backup plans or ideas with timelines
10. Ensuring that all information is present that needs to be
11. Making a timely decision
12. Communicating the decision made, how the decision was made, and who was involved in the decision-making process
13. Updates regarding the decision made
14. Transitional steps to the decision and how the business is affected by the decisions made

Good decisions may fail once implemented, so the rationale behind the decision is imperative for the people involved. The people affected by the decisions that are made are the most critical to the process and should be considered when coming up with conclusions for the decisions. Once important decisions are made through an intelligent thought process, the same methodology should be used in the future to make further decisions in an effective manner. The buy in for the decision-making process should also be shared so there is consensus on how the business makes decisions in the long run.

Decisions, whether large or small, impact everyone at hand. Because this is critical to communication methodologies, the impact of the decisions must be thought about thoroughly and the analyses behind decisions must be extremely intellectual and effective. Once the decisions made are effective and the communication has been thought through, individuals will begin trusting the process and making the changes habit.

A great deal of time, a lot of dialog occurs whether at a meeting or during a conversation, and there is perception that there was a decision made. It should be clear that dialog is not decision making. Ideas will slip away if not documented or not have action because nobody was assigned to be accountable for the idea. During the conclusion of a meeting or conversation, a decision should be made if needed by stating who is to do what or asking what needs to be done next. A list of action items should be concluded from meetings and conversations so that there is no lack of clarity. It should be made clear who is responsible for the decision made and why. As a leader, delegating decision making should occur or a decision should be made for others. What decision to be made should be clear along with timelines of the decision making. If there is no timeline, the decision will not end up being made again. If people really want a say in the decision, a vote should be taken so that others do not think only managers or higher-ups grant decisions. Make open, honest, and healthy decisions jointly, which can be agreed upon by all parties.

Decision making should always include the following questions:

- *Who cares about the decision?* Deciding who will be affected by the decision will make a difference as to who is contributing to the information.
- *Who has the most valid information?* Determining who has valid and informative expertise will help qualify what decision is made.
- *Who has to agree upon the decision?* If managers or higher ups need to be informed of the decision and may have a different say, they must be involved in the decision-making process.
- *How quickly can you make the decision?* Making a decision as quick as possible should involve utilizing the least amount of people who absolutely need to be involved while still making a decision that everyone will agree to. If enough people are involved to make a good decision, the decision will be made faster with more commitments.

As a leader, making decisions is part of the job, but people should feel as though the leader does not make the decisions they do just because they are the leader. People want to know that decisions were well thought out, flexible, and data behind the decision can be provided. If there are multiple options for the decision that are especially controversial, the people who will be impacted in the decision should be thought about intensely if not involved. Information will only help validate the decision-making process, so being prepared is key for the decision making in this situation. Ensure that if other people are involved in the decision making, their voice is actually heard. It does not make sense to involve people in the decision-making process and then not taking into consideration what they actually had to say. If a decision needs to be made without a vote, it should be explained why the decision had to be made to relieve controversy or drama. If a vote is made, but it is clear what decision should have been made, the situation will again involve controversy. In some cases, decisions need to be made without the vote because it is for the benefit of the business or a long-run outcome. Once the decision is made and announced, allowing postdecision lobbying must be eliminated. The decision should be made in the open and the data behind the decision should be stated. Questions may be asked about the decision, but allowing arguments only makes the decision-making process longer and more tedious.

If important stakeholders are unable to attend the decision-making meeting, a decision should still be made if the decision needs to be made in a timely manner. The most intelligent decision based on the available information should be made and the decision should still be firm. The decision should not be changed later just because one key stakeholder was not able to make the decision-making meeting. This process should be discussed with key stakeholders so they are aware of what will occur if they are not available for key decision-making meetings. If they must be involved, then the decision-making meeting must be postponed. Documenting the decision, the information behind making the decision, and the key players involved will help later when determining why a particular decision was made. This also minimizes decisions being changed frequently. The clarity of the decision must be made at the end of the day and people must be held accountable to the decisions made.

References

Covey, Stephen R. 2004. *The 7 Habits of Highly Effective People.* New York: Free Press.

Gebelein, Susan H., Brian Davis, Kristie J. Nelson-Neuhaus, Carl J. Skube, David G. Lee, Lisa A. Stevens, and Lowell W. Hellervik. 2004. *Successful Manager's Handbook.* Roswell, GA: PreVisor.

Patterson, Kerry, Joseph Grenny, Ron McMillan, and Al Switzler. 2002. *Crucial Conversations: Tools for Talking When Stakes are High.* New York: McGraw-Hill.

9

Visual Communication

Communication is the real work of leadership!

—Nitin Nohria

No matter the method of communication, it is still communication if it is performed functionally. Think about when you travel to a foreign country, what is the first thing you learn to do? Normally, learning the signals of major important items such as knowing where the restroom is, finding transportation, and asking about food and accommodations. The majority of the time this communication methodology is done through certain hand gestures because there is a language barrier. Language barrier or not, the communication is still performed and understood. This methodology is all still communication, just visual communication. The same aspect goes for the workplace; people aren't always available for knowing what happened in prior days, shifts, or during the same times at different locations. Graphs, signs, and visual controls take place for the communication.

Technology for visual communications can be quite simple. Lean and visual communications go hand in hand. Lean Six Sigma is a philosophy about reducing waste and being more efficient, and visual communications are performed in order to be more efficient and avoid wasting time. Visual communication helps employees to think differently and be more self-disciplined. A lot of times visual communication can lead to motivating employees because there is a normal competitive nature that is involved. If a graph is seen where there is a target and an actual where the actual is not met, automatically people want to excel to meet the target. Visual communications are an aid to increase uptime and quality while saving money, which are all done for the customers. Showing employees amounts of defects is a means to not deter them but to help them correct the problems and not make the same mistakes. Visual communications can help assist in the following areas in a factory or at a business:

- Increase efficiency
- Decrease defects
- Reduce Work in Progress (WIP)
- Increase productivity
- Decrease motion

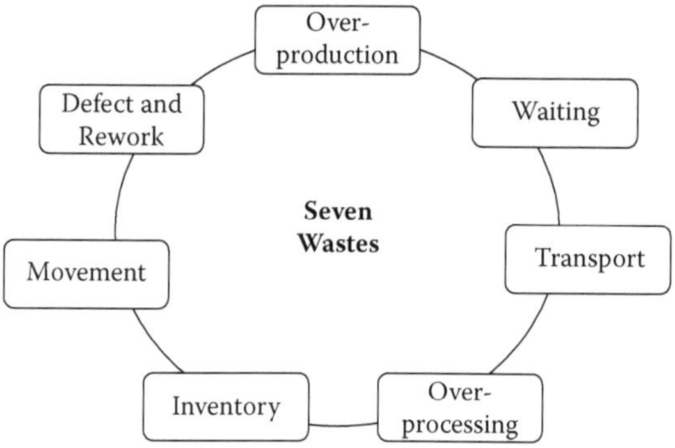

FIGURE 9.1
Seven wastes.

- Decrease batches and increase continuous flow systems
- Reduce costs
- Increase safety
- Increase customer satisfaction

The goal is to eliminate wastes (Figure 9.1).

Companies blame poor management or corporate when diminishing profits occur, but the root cause is from a lack of a lean system incorporating visual communication. Information not being portrayed is the biggest cause of problems because the communication aspect was lost. The Continuous Improvement aspects of increasing operational excellence while communicating the gains are important for a successful workplace. The improvement strategy used is what makes a company successful. Not only should a company have these measures in place, but it needs to ensure the sustainability of old improvements are taking place while incorporating new and Continuous Improvements daily.

Visual communication requires self-discipline and self-motivation. It takes an incorporated group for the fundamentals to work and cannot be successful with just one person. This initiative needs top management and must be driven to meet customer satisfaction. Visual standards and controls are a means of communication for all work aspects. They tell you when something needs to be done, where someone needs to be and when, how the progression of a project is going, and many safety measures.

It is important to know what to do with the communication and information given. Data is useless unless performance is changed due to the data given. Visual communication is the means to take data and make meaningful progressions to improve the workplace with sustainable performance metrics. Companies must grow with realistic expectations and then must be

TABLE 9.1
Production Capabilities

Line	Speed (lbs per hour)	Target (pounds)	Hours Needed per Week	Days Needed per Week
A	50	2500	50	3
B	100	3400	34	5
C	75	2200	29	6
D	85	6500	76	2

the best in the business in order not to lose the business. Gaining alignment will allow customers to grow. Some of these acts are quite simple.

An example of a factory:

Production Line A has a target of 2500 pounds to be made in week 1.

Production Line B has a target of 3400 pounds to be made in week 1.

Production Line C has a target of 2200 pounds to be made in week 1.

Production Line D has a target of 6500 pounds to be made in week 1.

The first process is to take the information and understand how long it will take to make each product. If Line A has a line speed of 50 pounds per hour, Line B has a line speed of 100 pounds per hour, Line C has a line speed of 75 pounds per hour, and Line D has a line speed of 85 pounds per hour, we need to first understand what this means in time. (See Table 9.1.) The bottleneck can be clearly seen as Line C since it takes the most days per week to complete. The second bottleneck is Line B. Therefore, staffing should be given to Lines C and B, respectively, first before moving to Lines A and D. A graph of the progress can easily be shown next as the production process continues over time (Figure 9.2). With this type of simple visualization, the

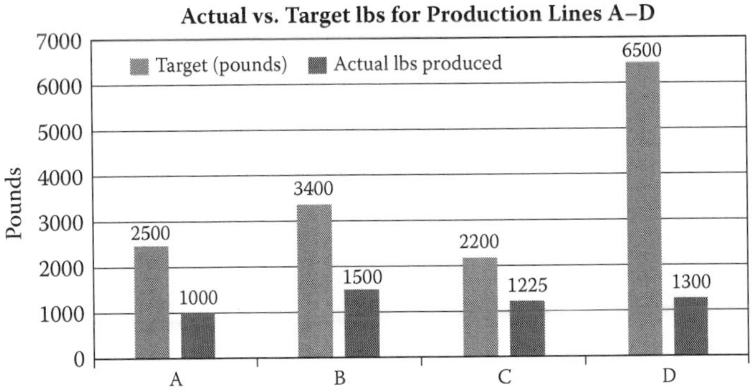

FIGURE 9.2
Actual versus target production Lines A to D.

progress can be monitored for both management and employees. It is obvious what work centers need to be the focus going forward and the progress can clearly be discussed to management.

It needs to be remembered that information is not progress, therefore the information given must lead to progress. Items need to be accomplished in order to improve rather than just communicating the information given. Knowing what information is present and what information to share is imperative for visual communication. Communicating too much unnecessary information will only lead to more chaos and confusion. The ease of the messages relayed need to be simplistic so that employees are not overwhelmed. Requirements of a well-maintained visual workplace setting include the following:

- Self-regulating areas
- Employees individually maintaining Continuous Improvement
- Self-explained methodologies
- Strategic concepts being incorporated into everyday work life

The same six common questions that we are used to should be asked to get to well-maintained workplace settings:

- Where?
- What?
- When?
- Who?
- How many?
- How?

As always, asking and telling why during situations answers questions and generates effective communication.

It is very important that when any of these means of communication questions are asked, that the answers are provided in polite, truthful, and complete terms. If an answer is unknown, it should be stated that the answer is not known and an investigation could be pursued to get the answer. Some answers may be confidential, and the confidentiality must be kept and stated as that. It is also important that all people are treated fairly and equally when these questions are asked so that there is no worry of favoritism.

Kaizen and 5S

Visual controls prevent masking of problems and allow errors to be seen immediately. 5S (sort, sweep, straighten, schedule, and sustain) comes into play with visual controls. Visual management utilizing 5S clears out rarely used items by sorting, organizing and placing them properly, straightening, cleaning work areas by sweeping, order by scheduling tasks and using preventive maintenance techniques, and using discipline to sustain progress. Visual controls improve value-added flows of processes by clearly demonstrating how work should be done and revealing errors if standard operating procedures (SOPs) are not followed. Well-designed process flows, graphs, charts, and other visual tools reflect improvement and motivate employees to continue improving.

Lean principles and other quality-driven initiatives are customer focused. Customer satisfaction is based on customer perceptions and expectations.

If a service level exceeds customer expectation, overprocessing has occurred. Customer perception has little or no correlation with expectations or actual service levels. Over time, customer expectations increase and overproduction decreases. Actual service levels normally remain fairly constant but customer perception fluctuates.

Kaizen methodologies are widely used for visual communication tools. *Kai* means to break apart or disassemble so that one can begin to understand. *Zen* means to improve. A Kaizen process focuses on an improvement objective by breaking the process into its basic elements to understand it, identify wastes, create improvement ideas, and eliminate the wastes identified. The basic philosophy of Kaizen is to manufacture products safely, when they are needed, and in proper quantities needed while reducing cycle and lead times.

Kaizen increases productivity, reduces WIP, eliminates defects, enhances capacity, increases flexibility, improves layout, and establishes visual management and measures. Kaizen increases productivity by revealing operator cycle time, eliminating waste, balancing workloads, utilizing value-added tasks, and producing to demand. Kaizen reduces WIP by determining needed and unneeded inventory.

Similar production outputs are grouped to balance production. Setup times can be reduced, and batches of outputs can be transported in smaller quantities. Preventive maintenance schedules can be established to aid consistent quality. Kaizen eliminates defects by asking why five or more times and reducing inventory so that improper manufacturing operations are quickly caught. Work is done under stable conditions and mistake-proof techniques are utilized. Kaizen enhances capacity and increases flexibility by finding and eliminating production bottlenecks caused by humans and machines.

Waste is identified and eliminated. Layout flexibility promotes the efficient flow of objects, people, and machines. Environments must be safe and clean, and allow regular preventive maintenance. Staffing is minimal and walking distances are shortened. Work should enter and exit an area at the same place.

Staff should communicate and the work balance should be even; workers should be able to help each other if needed.

Kaizen distinguishes equipment and operation improvements. Equipment-based improvements involve capital and time, may require major modifications, and may not produce cost savings. Operational-based improvements (a) change standard operating procedures; (b) change positions of layouts, tools, and equipment; (c) simplify tools by adding chutes, knock-out devices, and levers; (d) improve equipment efficiency without drastic modification; and (e) involve little cost and focus on cost reduction.

Kaizen focuses on a single piece of equipment or set of flows, synchronized movements, shortened transfer distances, movement of inventory into designated or finished states, and maintaining buffers between flows to keep flows from disturbing each other. Quality is always an issue for Kaizen, Lean, and Six Sigma. Defects are to be reduced, flows are to be improved, and processes must be streamlined. Finally, safety and environmental issues must be considered when operations and layouts are changed. Safety takes priority over any cost savings or productivity increases. Kaizen has 10 basic rules:

1. Think outside the box. No new idea should be considered a bad idea.
2. Determine how a task can be done, not how it cannot be done.
3. No excuses. Question current practices.
4. Perfection may not come immediately; improvements are required.
5. Mistakes should be corrected as soon as possible.
6. Ideas should be quick and simple, and not involve great amounts of money.
7. Find value in other people's ideas.
8. Ask why at least five times to find root causes.
9. Consult more than one person to find a true solution.
10. Kaizen ideas are long term.

Visual management and measures promote successful layouts. Visual management involves bins, cards, tags, signals, lights, alarms, and other signaling mechanisms. Visual systems include:

- Indicators such as signs, maps, and displays convey passive information.
- Signals such as alarms or lights are assertive devices.
- Controls provide aggressive information by monitoring size, weight, width, or length.
- Guarantees such as sensors, guides, and locators provide assured information.

Visual management systems cover tasks such as housekeeping via 5S, standard operating procedures, detecting errors and defects, and eliminating errors by mistake-proofing processes. Some of these techniques are known as methods, time measurements, or Maynard operations systems techniques.

How to Show Yields for Visualizing Progress

A *first pass yield* (FPY) indicates the number of good outputs from a first pass at a process or step. The formula is

$$FPY = \text{Number accepted}/\text{Number processed}$$

The formula for the first pass yield ratio is

$$\%\ FPY = (\text{Number accepted}/\text{Number processed}) \times 100$$

If 45 outputs are satisfactory among 60 produced, the FPY is 45/60 or 0.75. Thus

$$\%\ FPY = 0.75 \times 100 = 75\%$$

This number does not include rework of the rejected product.

Rolled throughput yield (RTY) covers an entire process. If a process involves three activities with FPYs of 0.90, 0.94, and 0.97, the RTY would be

$$0.90 \times 0.94 \times 0.97 = 0.82$$

Therefore

$$\%\ RTY = 0.82 \times 100 = 82\%$$

Continuous flow processing (CFP) is another aspect of operational Kaizen. It is characterized by continuous process flows, production paced according to Takt time (explained in further detail in Chapter 11, "Continuous Improvement Toolkit"), and pulling subsequent processes.

Takt time is a Kaizen tool used in the order-taking phase. *Takt* is a German word for pace. Takt time is defined as time per unit. This is the operational measurement to keep production on track. To calculate Takt time, the formula is Time available/Production required. Thus, if a required production is 100 units per day and 240 minutes are available, the Takt time is 240/100 or 2.4 minutes to keep the process on track. Individual cycle times should be balanced to the pace of Takt time. To determine the number of employees required, the formula is (Labor time/Unit)/Takt time. Takt in this case is time per unit. Takt requires visual controls and helps reduce accidents and injuries in the workplace. Monitoring inventory and production WIP will reduce waste or muda. *Muda* is a Japanese term for waste where waste is defined as any activity that consumes some type of resource but is non-value added for the customer. Customers are not willing to pay for this resource because it is not benefiting them. Types of muda include scrap; rework; defects; mistakes; and excess transport, handling, or movement.

To achieve CFP, machines must be arranged in the order of the process, one-piece flow of production must occur, multitasking must take place, easy moving must be possible, and U-cell layouts should be utilized. Pull production occurs when material is "pulled" from process to process only when needed. If a subsequent process already has the material, nothing is done.

*Kanbans a*re communication signals that control inventory levels while ensuring even and controlled production flow. Kanbans signal times to start, times to change setups, and times to supply parts. Kanbans work only if monitored consistently.

Value stream mapping reveals why excessive waste is introduced. The steps for value stream mapping are:

- Map current state processes
- Summarize current state processes
- Map future state processes
- Summarize future state processes
- Develop short-term plans to move from current to future state
- Develop long-term plans to move from current to future state
- Implement risks, failures, and processes for transitions
- Map key project owners, key dates, and future dates of future states
- Continue to map new future states when future states are met

The value stream adds value from a customer's view.

All steps in value stream mapping should deliver a product or service to customers.

A value stream involves multiple activities that convert inputs into outputs. All processes follow sequences of starts, lead times, and ends. The first goal is to reduce problematic areas by understanding the main causes for defects. The next step is to reduce lead time. The final step is to find the percentage of value-added time (VAT):

$$\%\ \text{VAT} = \text{Sum of activity times/Lead time} \times 100$$

When the sum of activity times equals lead time, the value-added time is 100%. For most processes, % VAT is 5% to 25%.

If the sum of activity times equals the lead time, the time value is not acceptable and activity times should be reduced. Lead time is the time from process start to end or time from receipt to delivery. Activity time is the time required to complete one output in an activity. Cycle time is the average time in which an activity yields an output, calculated as the activity time divided by the number of outputs produced in one cycle of an activity. Value-added definitions and results are:

- Customer value-added time (VAT)
  - Benefits customers and produces competitive advantage
  - Customers will pay for the tasks

- Non-value-added time (NVAT)
  - Customer wants to lower the prices by eliminating task
  - Task is eliminated or merged with upstream or downstream tasks

- Business value-added time (BVAT)
  - Required for business reasons such as conforming to laws or regulations
  - Required to reduce financial risks and liabilities
  - Required due to the nature of the process

Graphical Analysis

Graphical analyses are visual representations of tools that show meaningful key aspects of projects. These tools are commonly known as dotplots, histograms, normality plots, Pareto diagrams, second level Paretos (also known as stratification), boxplots, scatter plots, and marginal plots. The plotting of data is a key beginning step to any type of data analysis because it is a visual representation of the data.

If a particular manufacturing company wants to understand from where the majority of their defect costs are coming while trying to reduce those costs, the steps are as follows and can be described in more detail in Chapter 11, "Continuous Improvement Toolkit." Start with Table 9.2. Pareto

TABLE 9.2

Defect Costs Cheese Manufacturing

Largest Areas of Defects in Cheese Production Plant	Cost
Production Line A	$500,000
Production Line B	$400,000
Production Line C	$350,000
Production Line D	$125,000
Production Line E	$100,000

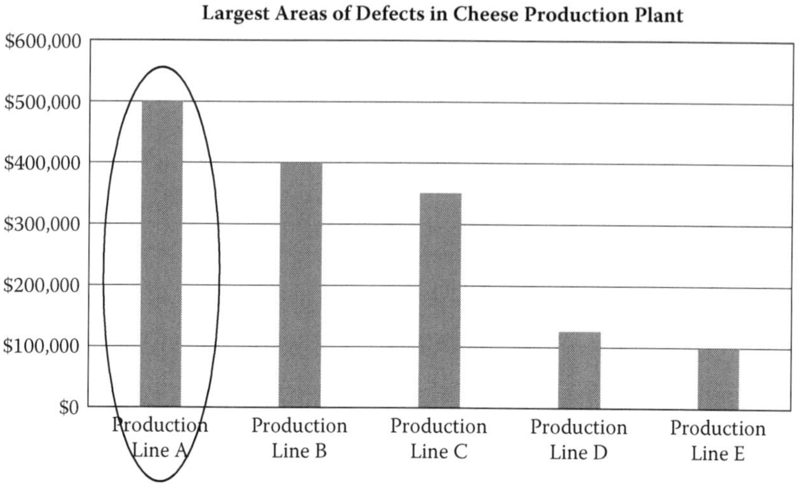

FIGURE 9.3
Pareto of manufacturing defect costs.

the costs to understand the largest hitters, as shown in Figure 9.3. The largest cost can be clearly seen for a next step in the process.

Map the Process

The importance of process mapping is to depict all functions in the process flow while understanding if the functions are value or non-value added. Any delays are to be eliminated and decisions are meant to be as efficient as possible. The purpose of process mapping is to have a visual image of the process.

Cause and Effect Diagram

After a process is mapped, the cause and effect (C&E) diagram can be completed. This process is so important because it completes root cause analysis. The basis behind root cause analysis is to ask why five times in order to get to the actual root cause. Many times, problems are "Band-Aided" to fix the top-level problem, but the actual problem itself is not addressed.

Fishbone Diagram

The fishbone diagram is broken out to the most important categories in an environment:

- Measurements
- Material
- Personnel
- Environment
- Methods
- Machines

This process requires a team to do a great deal of brainstorming where its focus is on the causes of the problems based on the categories. The "fish head" is the problem statement.

Failure Mode and Effect Analysis (FMEA)

In order to select action items from the C&E diagram and prioritize the projects, failure mode and effect analyses (FMEAs) are completed. The FMEA will identify the causes, assess risks, and determine further steps. The steps to an FMEA are the following:

1. Define process steps.
2. Define functions.
3. Define potential failure modes.
4. Define potential effects of failure.
5. Define the severity of a failure.
6. Define the potential mechanisms of failure.
7. Define current process controls.
8. Define the occurrence of failure.
9. Define current process control detection mechanisms.
10. Define the ease of detecting a failure.
11. Multiply severity, occurrence, and detection to calculate a risk priority number (RPN).
12. Define recommended actions.
13. Assign actions with key target dates to responsible personnel.
14. Revisit the process after actions have been taken to improve it.
15. Recalculate RPNs with the improvements.

Visual Management Boards and Techniques

Visual boards show team successes and give communication without every team member needing to be present. A visual management system normally consists of some type of a large whiteboard that has key points that are important to each team member. Common visual boards consist of the following points:

- Each person writes down three key items they are working on and what their holdup is
- Recognition of a key associate
- Capital spending box
- Reminders
- Calendar reference
- Meeting time
- Overdue action items

It is important to meet once a week at the same time to review the visual management board. Each person is accountable for their box and should fill it out before the meeting.

Utilizing different colors for certain topics, such as pink for important items, helps people visually separate items. Suggestions from the visual management boards should be taken, because the board should be always changing and improving. It is also important to remember that people react to colors, shapes, and so forth, so they should be utilized. Examples can be seen in Figure 9.4.

FIGURE 9.4
Traffic light visual.

FIGURE 9.5
Measuring cup visual.

Types of measurements that can be seen visually are also key indicators of whether or not a process is performing well. A visual technique for this is shown in a simple measuring cup in Figure 9.5. The purpose of these visual controls is to show what is right, what is wrong, what is complete, where the path going forward should be, what step to take next, who to see, and common standard operating techniques. These techniques tell us at a glance how things are going and points out imperfections or abnormal conditions instantaneously. Visual techniques will enhance performance to the next level improving office and factory performances (Figure 9.6).

A visual management board helps create lean management (Figure 9.7) by doing the following: keeping track of projects is also critical to ensuring they are completed, people are held accountable, and dates are prioritized. A simple spreadsheet laying out these characteristics will help keep teams on task (Figure 9.8). The following items should be included on this prioritization spreadsheet:

- Area for Improvement
    - Labor
    - Waste
    - Efficiency

- Priority
    - 1—Active
    - 2—Will be active within 1 month
    - 3—Will require resources

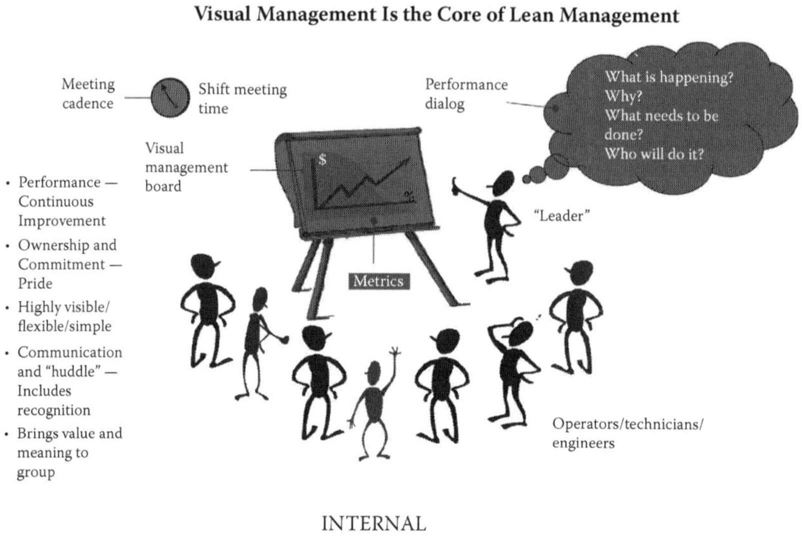

FIGURE 9.6
Visual Management and Lean Management.

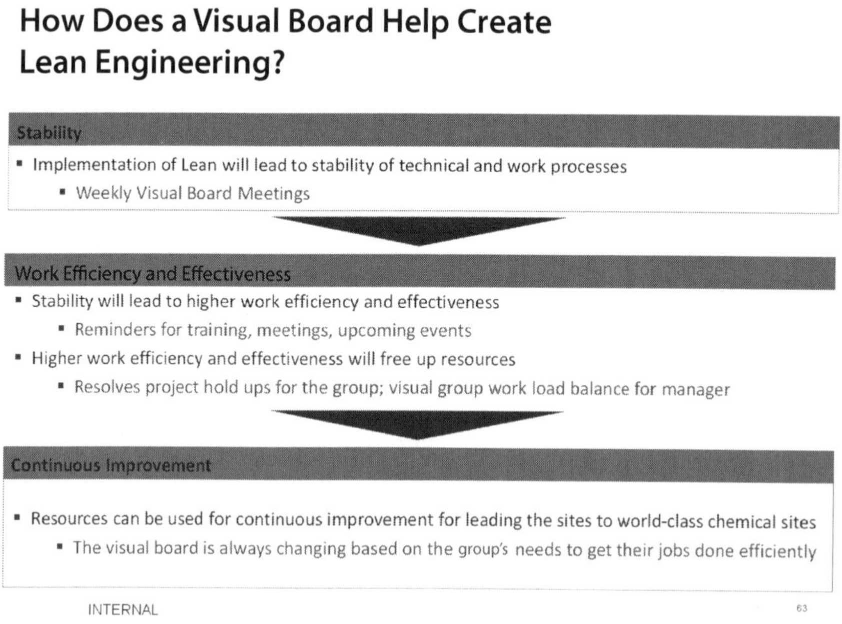
How Does a Visual Board Help Create Lean Engineering?
Stability
▪ Implementation of Lean will lead to stability of technical and work processes
▪ Weekly Visual Board Meetings
Work Efficiency and Effectiveness
▪ Stability will lead to higher work efficiency and effectiveness
▪ Reminders for training, meetings, upcoming events
▪ Higher work efficiency and effectiveness will free up resources
▪ Resolves project hold ups for the group; visual group work load balance for manager
Continuous Improvement
▪ Resources can be used for continuous improvement for leading the sites to world-class chemical sites
▪ The visual board is always changing based on the group's needs to get their jobs done efficiently
INTERNAL
63

FIGURE 9.7
Visual Management and Lean Management Engineering.

Area	Priority	Department	Project	Action	Follow-Up	Completion	Savings	Responsible	Due Date	Cross-Functional Team	Notes
Labor	1					0					
Waste	1					25					
Labor	1					0					

FIGURE 9.8
Prioritization spreadsheet.

- Department
- Project
- Action
- Follow Up
- Completion Percentage
- Savings
- Person Responsible
- Due Date
- Cross-Functional Team
- Notes

A lean enterprise consists of all processes combined. Ensure that all processes are understandable to all, goal driven (SMART), quick visuals, and standardized, that way the results are shown over time versus waiting to the last minute to show results. Targets should always be given, so the visuals are visible, flexible, and simple to make action items out of to improve. Motions need to be eliminated, metrics need to be displayed and standardized, and employees need to be strategic thinkers in a self-driven environment. A lean enterprise is market driven; when customers cannot get something from one supplier or are not happy with that supplier, they simply go elsewhere, also known as a competitor.

References

Agustiady, Tina. 2012. *Sustainability: Utilizing Lean Six Sigma Techniques*. Boca Raton, FL: Taylor & Francis/CRC Press.

Galsworth, Gwendolyn. 2005. *Visual Workplace, Visual Thinking*. Portland, OR: Visual-Lean Enterprise Press.

Kovach, Tina. 2012. *Statistical Techniques for Project Control*. Boca Raton, FL: Taylor & Francis/CRC Press.

10

What Is Lean Six Sigma and TPM?

When you are finished changing, you're finished.

—Benjamin Franklin

Chapter 11, "Continuous Improvement Toolkit," provides a list of all of the tools utilized for Lean, Six Sigma, and TPM (Total Production Maintenance). The following is a brief overview of what Lean, Six Sigma, and TPM are.

Lean

Lean is a terminology that is well known and defined as an elimination of waste in operations through managerial principles. Many principles comprise the Lean concept, but the major thought to remember is effective utilization of resources and time in order to achieve higher quality products and ensure customer satisfaction. Remember, defects are anything that the customer is unhappy with and is a term utilized in Six Sigma. Six Sigma identifies and eliminates these defects so that the customer in turn is satisfied. Customers are the number one focus and if they are unhappy, they will have no problem going elsewhere, which most likely is a competition for the business. Coupling Lean and Six Sigma will reduce waste and reduce defects. The concept will be called Lean Six Sigma going further.

The most basic concept when discussing waste reduction begins with Kaizen. Kaizen is a Japanese concept defined as "taking apart and making better." The concept takes a vast amount of project management techniques to facilitate the process going forward. 5S processes are the most predominant and commonly known for Kaizen events.

There are two main concepts of Lean:

- Waste
- Visual Management

Lean uses five practices:

- Focus
- Common approach

- Full employee engagement
- Sharing of best practices
- Links to company goals

No two Lean implementation plans are the same. They require support processes and strong tools. Piloting initiatives show where mistakes are prevalent firsthand and where improvements can be made.

Six Sigma

Six Sigma is best defined as a business process improvement approach that seeks to find and eliminate causes of defects and errors, reduce cycle times, reduce costs of operations, improve productivity, meet customer expectations, achieve higher asset utilization, and improve return on investment (ROI). Six Sigma deals with producing data driven results through management support of the initiatives. Six Sigma pertains to sustainability because without the actual data, decisions would be made on trial and error. Sustainable environments require having actual data to back up decisions so that methods are used that will have improvements for future generations. The basic methodology of Six Sigma includes a five-step method approach that consists of the following:

1. *Define*—Initiate the project, describe the specific problem, identify the project's goals and scope, and define key customers and their Critical to Quality (CTQ) attributes.
2. *Measure*—Understand the data and processes with a view to specifications needed for meeting customer requirements, develop and evaluate measurement systems, and measure current process performance.
3. *Analyze*—Identify potential cause of problems; analyze current processes; identify relationships between inputs, processes, and outputs; and carry out data analysis.
4. *Improve*—Generate solutions based on root causes and data-driven analysis while implementing effective measures.
5. *Control*—Finalize control systems and verify long-term capabilities for sustainable and long-term success.

The goal for Six Sigma is to strive for perfection by reducing variation and meeting customer demands. The customer is known to make specifications for processes. Statistically speaking, Six Sigma is a process that produces 3.4 defects per million opportunities. A defect is defined as any event that is outside of the customer's specifications. The opportunities are considered

TABLE 10.1
Six Sigma Defects per Million Opportunities

Sigma Spread	DPMO	Percent Defective	Percent Yield	Short-Term C_{pk}	Long-Term C_{pk}
1	691,462.00	69%	31%	0.33	–0.17
2	308,538.00	31%	69%	0.67	0.17
3	66,807.00	7%	93.30%	1	0.5
4	6,210.00	0.62%	99.38%	1.33	0.83
5	233	0.02%	99.38%	1.67	1.17
6	3.4	0%	100%	2	1.5

any of the total number of chances for a defect to occur. Table 10.1 explains the defects per million opportunities and sigma levels.

The normal distribution that underlies the statistical models of the Six Sigma model is shown in Figure 10.1. The Greek letter σ (sigma) marks the distance on the horizontal axis between the mean μ and the curve inflection point. The greater the distance, the greater is the spread of values encountered. The figure shows a mean of 0 and a standard deviation of 1, that is, $\mu = 0$ and $\sigma = 1$. The plot also illustrates the areas under the normal curve within different ranges around the mean. The upper and lower specification limits (USL and LSL) are ±3σ from the mean or within a six-sigma spread. Because of the properties of the normal distribution, values lying as far away as ±6σ from the mean are rare because most data points (99.73%) are within ±3σ from the mean except for processes that are seriously out of control.

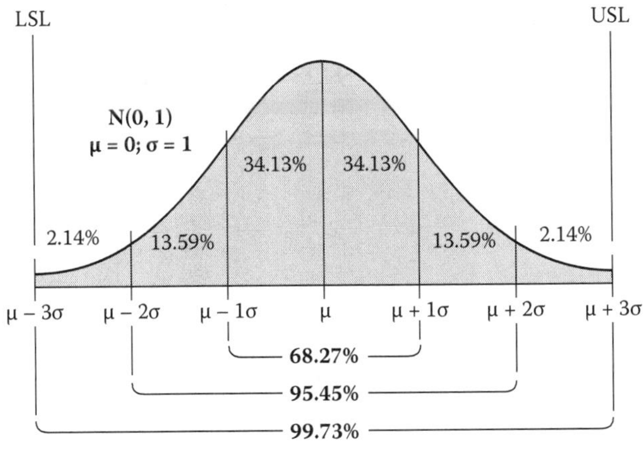

FIGURE 10.1
Areas under the normal curve.

Six Sigma allows no more than 3.4 defects per million parts manufactured or 3.4 errors per million activities in a service operation. To appreciate the effect of Six Sigma, consider a process that is 99% perfect (10,000 defects per million parts). Six Sigma requires the process to be 99.99966% perfect to produce only 3.4 defects per million, that is, 3.4/1,000,000 = 0.0000034 = 0.00034%. That means that the area under the normal curve within $\pm 6\sigma$ is 99.99966% with a defect area of 0.00034%.

TPM

TPM has been a well-known activity that has several names associated within. Many people associate TPM with Total Predictive Maintenance or Total Preventative Maintenance. The association explained next will be Total Productive Maintenance, but includes the aforementioned as well.

TPM originated in Japan in 1971 as a methodology to improve machine availability and throughput through the utilization of more efficient maintenance and production resources.

The goal of TPM is to increase job satisfaction through the following means:

- Reduced breakdowns
- Reduced quality issues
- Reduced safety/environmental incidents
- Reduced costs
- Improved throughput
- Competitive advantage
- Emergency and unplanned maintenance at a minimum

There are four main objectives of TPM:

1. Avoid waste in quickly changing environments
2. Reduce costs of manufacturing
3. Produce a low batch quantity at the earliest possible time
4. Goods sent to customers must be "nondefective"

TPM is also known for being compared to pillars inside of a house. The house and pillars consist of the following (see Figure 10.2).

Pillar 1: 5S

- TPM starts with 5S (Figure 10.3); issues cannot be seen clearly in an unorganized place.

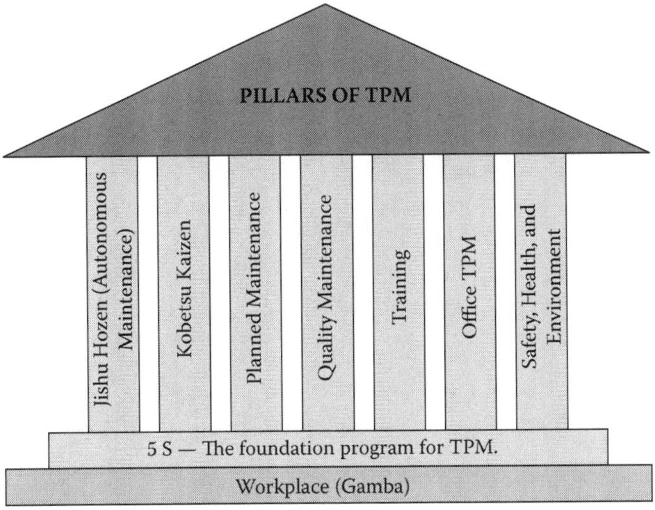

FIGURE 10.2
Pillars of TPM.

- Cleaning and organizing will uncover problems.
- Making problems visible is the first step of improvement.
  - Sort
  - Straighten
  - Shine/sweep
  - Standardize
  - Sustain

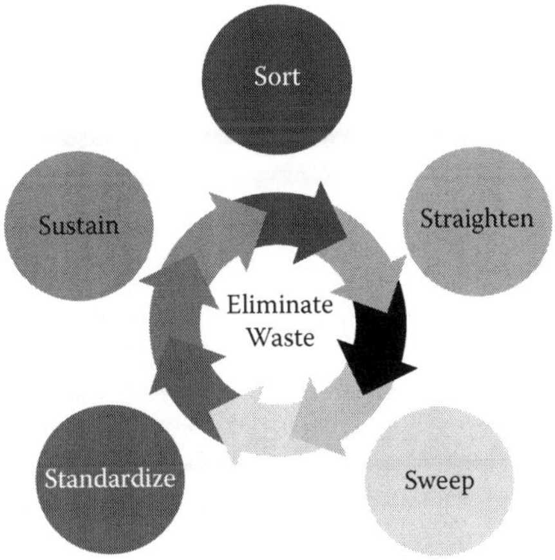

FIGURE 10.3
5S.

FIGURE 10.4
Autonomous maintenance (Jishu Hozen).

Pillar 2: Autonomous Maintenance (Jishu Hozen) (Figure 10.4)

- Empowering and developing operators to be able to take care of small maintenance tasks
- Frees up skilled maintenance people to spend time on more value added activity and technical repairs
- Operators are responsible for upkeep of their equipment to prevent it from deteriorating

Targets for Autonomous Maintenance

- Reduce process time by *X*%
- Increase autonomous maintenance (AM)
- Operations of equipment is uninterrupted
- Operators are flexible and maintain other equipment
- Defects are eliminated at the source through employee participation

Steps

1. Preparation of employees
2. Initial cleanup of machines
3. Take countermeasures
4. Fix tentative Jishu Hozen standards
5. General inspection
6. Autonomous inspection
7. Standardization
8. Autonomous management

Kaizen

改善

To make better

FIGURE 10.5
Kaizen Symbol.

Pillar 3: Kobetsu Kaizen (Particular Case)

- Japanese term (Figure 10.5)
- *Kai* means "change"
- *Zen* means "for the better"
- Kaizen = Continuous Improvement

Concept

- Small incremental improvements
- Improvements add up over time

Targets

- Zero losses sustained with minor stops, measurements, and adjustments
- Zero defects and unavoidable downtime
- Reduce manufacturing costs by X%

Steps

1. Practice concept of Zero Losses in every sphere of activities
2. Relentless pursuit to achieve cost reduction targets in all sources
3. Relentless pursuit to improve overall plant equipment effectiveness
4. Extensive use of planned maintenance analysis as a tool to eliminate losses
5. Focus on easy handling of operators

Pillar 4: Planned Maintenance

- Aimed to have trouble-free machinery and equipment with zero defects for 100% customer satisfaction
- Become proactive versus reactive while utilizing trained maintenance staff to help train operators to better maintain their equipment

Targets

- Zero equipment failure and breakdown
- Improve reliability and maintainability by 50%

- Reduce maintenance costs by 20%
- Ensure availability of spares at all times

Steps

1. Equipment evaluation and recording present status
2. Restore deterioration and improve weakness
3. Building up an information management system
4. Prepare a time-based information system, select equipment, parts, and members, and map out a plan
5. Prepare predictive maintenance system by introducing equipment diagnostic techniques
6. Evaluation of planned maintenance

Pillar 5: Quality Maintenance

- Aimed toward customer delight through highest quality through defect-free manufacturing
- Focus is on eliminating nonconformances in a systematic manner
- We gain understanding of what parts of the equipment affect product quality and begin to eliminate current quality concerns then move to potential quality concerns
- Transition is from reactive to proactive

Targets

- Zero customer complaints
- Reduce in-process defects by 50%
- Reduce cost of quality by 50%

Quality defects are classified as *customer-end defects* and *in-house defects*. For customer-end data, we get data on:

1. Customer end-line rejection
2. Field complaints
3. In-house data include data related to products and data related to process

Pillar 6: Training

- Aimed to have multiskilled and energized employees who have high morale and are eager to come to work to perform all their required functions independently and effectively
- Education is given to operators to upgrade their skills
- Employees should be taught to achieve the skills of training through the different phases (Figure 10.6):
  - Phase 1—Do not know
  - Phase 2—Know the theory but cannot do

FIGURE 10.6
Training Demonstration.

- Phase 3—Can do but cannot teach
- Phase 4—Can do and also teach

Targets

- Achieve and sustain downtime at zero on critical machines
- Achieve and sustain zero losses due to lack of knowledge/skills/techniques
- Aim for 100% participation in suggestion scheme

Steps

1. Setting policies and priorities, and checking present status of education and training
2. Establishment of training system for operations and maintenance skills
3. Training the employees for the operation and maintenance skills
4. Preparation of training calendar
5. Kickoff the training
6. Evaluation of activities and study of future approach

Pillar 7: Office TPM

- Office TPM should be started after activating from other pillars of TPM (Jishu Hozen, Kobetsu Kaizen, and Quality Maintenance/Planned Maintenance)
- Office TPM must improve productivity, efficiency, and flow in the administrative functions while identifying losses
- Analysis of processes and procedures toward office automation is sought after

12 major losses are covered:

1. Processing loss
2. Cost loss including in areas such as procurement and accounts marketing leading to high inventories
3. Communication loss
4. Idle loss
5. Setup loss
6. Accuracy loss
7. Office equipment breakdown
8. Communication channel breakdown
9. Time spent on retrieval of information
10. Nonavailability of correct online stock status
11. Customer complaints due to logistics
12. Expense on emergency dispatches/purchases

Pillar 8: Safety, Health, and Environment

- Focus to create a safe workplace and a surrounding area that is not damaged by process or procedures
- This pillar will play an active role in each of the other pillars on a regular basis
- Zero accident
- Zero health damage
- Zero fires (Figure 10.7)

There are different types of maintenance involved with TPM.

1. Breakdown maintenance—You wait until the machine completely fails, then you repair it to working order. Usually assigned to machines with backups in place or machines that are of low importance to production.
2. Preventive maintenance—You maintain the machine, while still in operation to prolong or prevent the machine's failure. This

FIGURE 10.7
Zero mindset mentality.

should be applied to machines of importance and high value to production.

   a. Periodic maintenance—A set schedule of maintenance routines assigned to prolong the life of the machine.
   b. Predictive maintenance—You predict when the machine may require maintenance based on analyzing past history, and apply maintenance just before the machine has failed in the past, extending its service life. Often applied in conjunction with preventative maintenance.

3. Corrective maintenance—A form of system maintenance that is performed after a fault or problem emerges in a system, with the goal of restoring operability to the system. In some cases it can be impossible to predict or prevent a failure, making corrective maintenance the only option.
4. Maintenance prevention—Machine engineering and design that is based on preventing the need for maintenance or for ease of access to machine parts so that maintenance may be carried out easily. Different than preventative maintenance in that the maintenance is performed while a machine is still in working order to keep it from breaking down. Preventive maintenance includes lubricating, tightening, and replacing worn parts.

Preventive maintenance is work that is done on a machine, often involving testing and the replacement of worn but still functioning parts, to prevent a failure (as against fixing something only when it breaks). Maintenance prevention means stopping maintenance from taking place. If used in a positive context: steps taken perhaps in the design of a device to make it require less maintenance.

Introducing TPM at a production unit has four main steps:

Step 1—Preparatory Stage

- Announcement by top management to all about TPM introduction in the organization
- Initial education and publicity for TPM
- Setting up TPM and Departmental committees
- Establishing the TPM working systems and target
- A master plan for institutionalizing

Step 2—Introduction Stage

Step 3—Implementation

Step 4—Institutionalizing Stage

Common Targets for Lean, Six Sigma, and TPM

Different groups in workplaces will have different targets. Some of the common targets are shown next for the various groups:

- Production
  - Obtain minimum 80% overall production efficiency
  - Obtain minimum 90% overall equipment effectiveness
  - Run the machine during lunchtime
- Quality
  - Zero customer complaints
- Cost
  - Reduce manufacturing costs
- Delivery
  - 100% success rate defined by customer
- Safety
  - 0 accidents
- Cross-functional teams
  - Multiskilled workers
  - Flexible workforce

Benefits of Lean, Six Sigma, and TPM

- Increased reliability and availability
- Reduced safety and environmental incidents
- Efficiency improvements/workforce utilization
- Increased ownership
- Personnel satisfaction

TPM is a critical adjunct to Lean manufacturing. If machine reliability or uptime of the machine is not predictable and not able to be sustained, the process must keep extra stocks to buffer against this uncertainty and flow through the process will be interrupted. Unreliable uptime is caused by breakdowns or badly performed maintenance. Correcting maintenance will allow uptime to improve and speed up production through a given area allowing a machine to run at its designed capacity of production.

Lean, Six Sigma, and TPM is a journey for educating and training the workforce to be familiar with machinery, parts, processes, efficiencies, losses, and damages while being productive.

References

Agustiady, Tina, and Adedeji B. Badiru. 2012. *Statistical Techniques for Project Control.* Boca Raton, FL: Taylor & Francis/CRC Press.

Agustiady, Tina, and Adedeji B. Badiru. 2012. *Sustainability: Utilizing Lean Six Sigma Techniques*. Boca Raton, FL: Taylor & Francis/CRC Press.

11

Continuous Improvement Toolkit

The Continuous Improvement Toolkit is an easy reference for what tool to use and when, and how to effectively teach the tools to employees who are not necessarily engineers. The implementation of the actual tools will also be taught in this chapter. The Continuous Improvement Toolkit will consist of the following topics that will be taught how to be used with real-life examples:

5S
7/8 Wastes
Kaizen
Fishbone Diagrams
Root Cause Analysis
Process Mapping
Financial Justification
One Point Lessons
Value Stream Mapping
Plan–Do–Check–Act
Poka Yokes
Kanbans
Pull and Push Flows
Visual Management
Cellular Processing
7/8 Wastes
Spaghetti Diagrams
Histograms
Pareto Charts
Capability Analysis
Control Charts
Defects per Million Opportunities
Project Charters
Supplier–Input–Process–Output–Customer (SIPOC)
Kano Model

Critical to Quality (CTQ)
Affinity Diagram
Measurement Systems Analysis
Gage R&R
Process Capabilities
Variation
Graphical Analysis
Cause and Effect Diagram
Failure Mode and Effect Analysis (FMEA)
Hypothesis Testing
Analysis of Variance (ANOVA)
Correlation
Simple Linear Regression
Theory of Constraints
Single Minute Exchange of Dies (SMED)
Total Productive Maintenance (TPM)
Design for Six Sigma (DFSS)
Quality Function Deployment (QFD)
Design of Experiments (DOE)
Mood's Median Test
Control Plans

5S

5S is a fundamental tool that serves as the foundation for many tools such as Lean, Six Sigma, TPM (Total Productive Maintenance), and Waste Management. Future improvements can be made once 5S is implemented. A clean workplace facilitates change where problems naturally stand out. There will also be more room for extra business while ensuring product quality and safety in a well-maintained environment. The goal of 5S is to be able to expose problems that prevent us from being successful in the future.

An unclean environment contains hidden risks for workers and equipment. Advantages of a 5S organization include time management, safety, quality, customer responsiveness, and visual controls. 5S principles will improve work environments where each S progresses through implementation.

5S stands for the following:

Sort—Identify necessary items, and eliminate and dispose of unneeded materials that do not belong in an area. This reduces waste, creates a safer work area, opens space, and helps visualize processes. It is important to sort through the entire area. The removal of items should be discussed with all personnel involved. Items that cannot be immediately removed should be tagged for subsequent removal.

Sweep—Clean the area so that it looks like new and clean it continuously. Sweeping prevents an area from getting dirty in the first place and eliminates further cleaning. A clean workplace indicates high standards of quality and good process controls. Sweeping should eliminate dirt, build pride in work areas, and build value in equipment.

Straighten—Have a place for everything and everything in its place. Arranging all necessary items is the first step. It shows what items are required and what items are not in place. Straightening aids efficiency; items can be found more quickly and employees travel shorter distances. Items that are used together should be kept together. Labels, floor markings, signs, tape, and shadowed outlines can be used to identify materials. Shared items can be kept at a central location to eliminate purchasing more than needed.

Standardize—Assign responsibilities and due dates to actions while using best practices throughout the workplace. All departments must follow standardized rules to comply with 5S. Items are returned where they belong and routine cleaning eliminates the need for special cleaning projects. Audits to clear up unnecessary items, organize items in designated places, and cleaning is part of the standardize phase. Anything out of place or dirty should be noticed immediately.

Sustain—Establish ways to ensure maintenance of manufacturing or process improvements. Sustaining maintains discipline. Utilizing proper processes will eventually become routine. Training is key to sustaining the effort and involvement of all parties. Management must mandate the commitment to housekeeping for this process to be successful.

If 5S is not implemented, waste may result. The seven main types of waste results (also see Figure 11.1) include:

- Overproduction
- Waiting
- Transport
- Overprocessing
- Inventory
- Movement

Sort—Segregate what is needed and what is not needed, and what is needed later on. Utilize red tags to determine which items are no

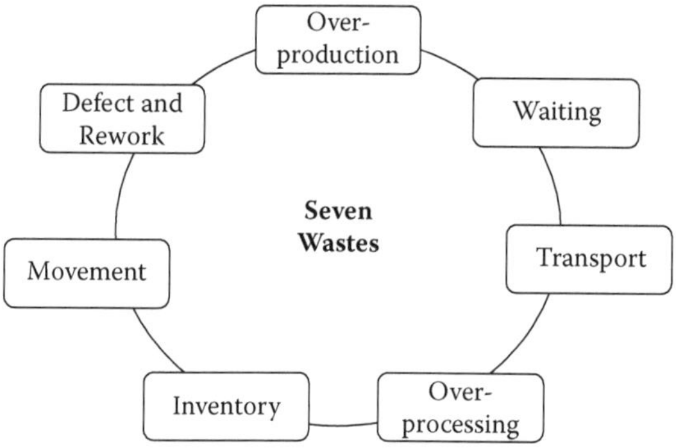

FIGURE 11.1
7 wastes.

longer needed. Discard what is not needed and segregate the needed items by the frequency in which they are used.

Red tags denote items that are not needed (Figure 11.2). Discard items that are not needed or move them to a storage area. Ensure items that are needed and items that might be needed are differentiated. Move items not needed on a weekly and daily basis. These tags must be managed weekly and everyone should be able to have a chance to respond to the red tags. Be critical when looking at these items. Eliminate defective items, store any excess items, and dispose of any obsolete stock.

Sweep—During the sweep phase, items are cleaned up and sometimes deep cleaned. The challenge is to implement regular housekeeping activities so anything dirty is automatically seen and made quite obvious. During the sweep phase, undesired dirt or debris is removed,

FIGURE 11.2
Red tag.

and a healthier and safer work environment is promoted. The customers also like the impression of a clean work environment. During this clean phase, inspections are also performed. The items to look for are faulty parts, anything hidden or broken, loose parts or parts in need of calibration, and any items that need to be refilled. The cleaning must be done on a regular basis. Specific tasks should be assigned to specific personnel. Ensure the proper supplies are close by for the cleaning to take place. A designated time should be set aside per day to do the cleaning as well, this is normally toward the end of the shift.

Straighten—Straightening includes a high level of organization. Arrangement of items should be placed so they are easy to find and can be readily retrieved. It should also be obvious if an item is missing. This phase should be known as "A place for everything, and everything in its place."

Visual management is a key to this phase to show where, what, and how many. This promotes ready retrieval and ready return. It is also important to remember that people react to colors, shapes, and so forth, so they should be utilized. Examples can be seen in Figure 11.3. Figure 11.4 and Figure 11.5 show the before and after results, respectively, of the Straighten phase.

Standardize—Communication plays a large role during the standardize phase. Best practices of one area must be communicated along with expectations of other areas. It is important that everyone is aware of their roles and responsibilities. Training plays a key role in this phase as well so that everyone is trained the exact same way. Continuous Improvement must be part of the phase as well so that best practices

FIGURE 11.3
Traffic light visual.

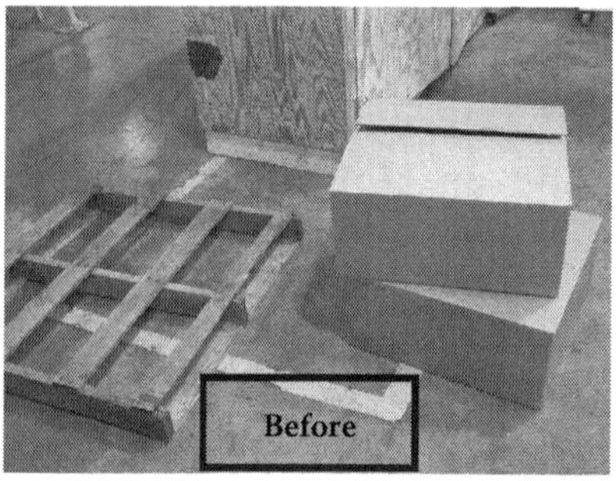

FIGURE 11.4
Before straighten.

continue to be used. Leaders will set the tone for the best practices to be used. They should gain commitment from employees through their motivation.

Sustain—Celebrating is the best way to motivate employees and ensure they keep on track with their improvements. Recognition is a vital portion of 5S to ensure high standards were met and agreed upon. It is important to sustain results so that the "old" way of doing things does not occur. Reinforcement of the program will ensure its success and convince other workgroups to go along the same path. Ensure logbooks are used for items that were discarded for future use. Pictures should also be taken from the exact same location to show before and after improvements (Figure 11.6). Action plans for the future must be made for the Sustain phase including audits to take place and the responsibilities of employees. Continuous Improvement again should be utilized to always improve the benefits of 5S.

FIGURE 11.5
After straighten.

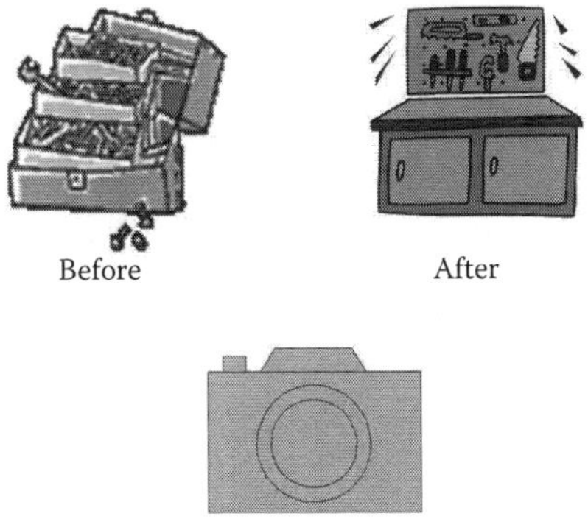

FIGURE 11.6
Before and after picture—5S.

7/8 Wastes

The 7 Wastes (Figure 11.7) are known as the following:

1. *Overproduction*—Producing more than is needed before it is needed
2. *Waiting*—Any nonwork time waiting for tools, personnel, parts, and so forth
3. *Transport*—Wasted efforts used to transport materials, parts, goods, etc., in or out of storage or between processes
4. *Overprocessing*—Performing more work than is needed
5. *Inventory*—Excess parts, finished goods, raw materials, and so on that are not being utilized

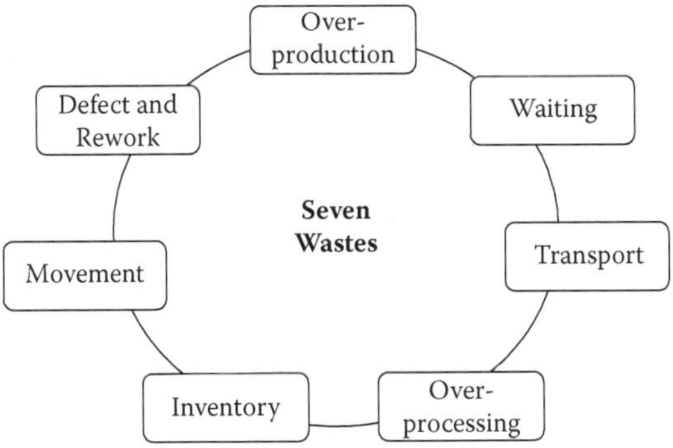

FIGURE 11.7
7 wastes diagram.

6. *Movement*—Wasted movements to move or store parts, or excess walking performed
7. *Defect and rework*—Defective products or parts

An eighth waste has been added as the following:

8. *Underutilization of employees*—Not fully utilizing an employee for a job

Kaizen

Kaizen is a Japanese word where *kai* means "change." *Zen* means "for the better." Kaizen equals Continuous Improvement. The concept includes small incremental improvements that will add up over time. It is important to realize that Kaizen does not necessarily mean a blitz or an event, even though an event or blitz often occurs for a Kaizen kickoff.

Kaizen is a Continuous Improvement event with the following:

- Dedicated resources
- Specific goals and deliverables
- Short time frame

It can also be known as:

- Rapid Continuous Improvement Event
- Kaizen Blitz
- Lean Event

Kaizen creates a vision of what our production system and manufacturing techniques should be while carrying out that vision by breaking through the status quo (Figure 11.8). The basic rules for Kaizen are the following:

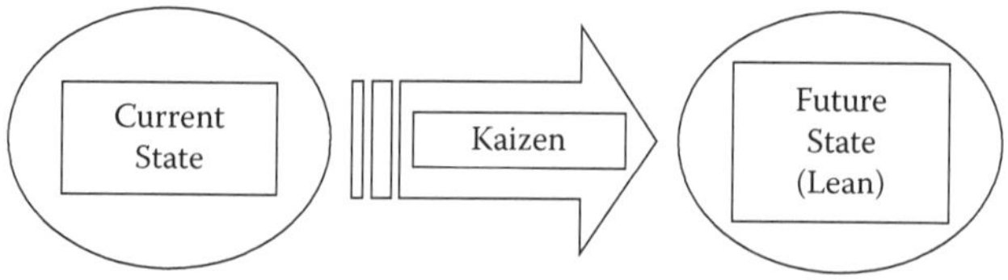

FIGURE 11.8
Current state to future state for Kaizen.

- Be open minded for changes
- There are no dumb questions or ideas
- Be positive
- Avoid spending money
- Question current practices and challenge the status quo
- Think about how to change versus how not to change
- Go, See, Think, Do
- Have fun

Kaizen drives improvements, which lead to a leaner business operating system.

Fishbone Diagrams

Fishbone diagrams are also known as cause and effect diagrams, which are used to understand knowledge about a process or a product. The diagrams were named after Dr. Kaoru Ishikawa, so they are sometimes known as Ishikawa diagrams as well. A team would come together to have a structured brainstorming revolving around the process or product. The brainstorming is emphasized in a graphical representation for visual management purposes. It serves as a communication tool. The cause and effect diagram assists in reaching a common understanding of the problem and exposes the potential drivers of the problem. Normally, 5Ms are utilized for this brainstorming. The 5Ms are generally manpower (personnel is more appropriate), methods, materials, machinery, and measurements. A sixth element of environment is frequently used as well. The diagram looks like a fish in that the head will be a box that describes the effect and the body will have bones that include the 5Ms (see Figure 11.9). Once each M is defined, a cause and a reason for the cause is identified for the specific M associated within. This type of brainstorming normally uncovers potential issues and results in many action items.

Root Cause Analysis

Special cause variation must be reduced with root cause analysis. This means to not Band-Aid these problems, but understand why these problems happened in the first place. Root cause analysis is the process of finding and

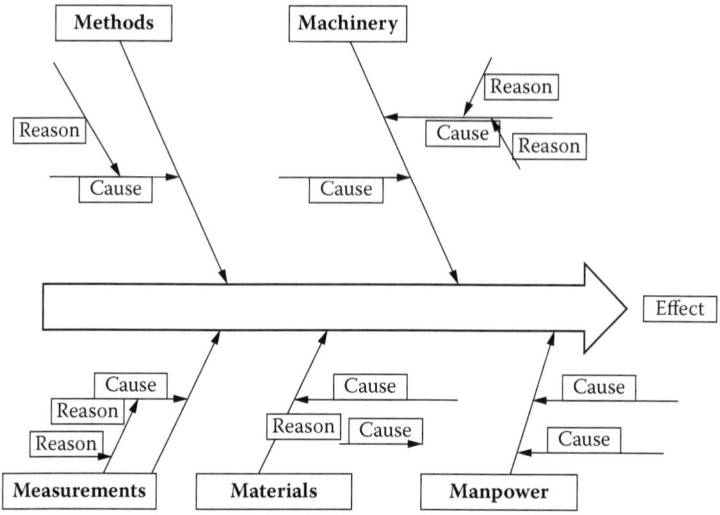

FIGURE 11.9
Fishbone diagram.

eliminating the cause, which would prevent the problem from returning. Only when the root cause is identified and eliminated can the problem be solved. Root cause analysis uses the DMAIC philosophy as a guide. The steps to root cause analysis are as follows:

1. Define the problem
2. Map the process
3. Gather data
4. Seek for root causes through fishbone diagrams/cause effect diagrams
5. Verify root causes with data
6. Develop solutions and prevention steps including costs and benefits
7. Pilot implementation plans
8. Implement
9. Control utilizing a monitoring plan and process metrics
10. Identify lessons learned

The steps are graphically shown in Figures 11.10 to 11.15.

Root Cause Analysis	Page 1 of 5
This form is a tool for finding causes of process problems and developing actions for eliminating, preventing, and minimizing future problems.	Date: 1/1/13

Incident Date: 1/1/13	RCA Initiated by: T Agustiady
Investigator: T Agustiady	
1.Define the problem	Describe the incident. What was defect, how many, how often, etc.

Line 1 incurs an average of one hour of down time per day on a ten hour shift. Eliminate 50% of mechanical downtime on line by 4/1/13 with Root Cause Analysis

Step (not all required, depending on problem)		Date completed
D	1. Define the problem	
	2. Map Process (if required)	
M	3. Gather data	
	4. Cause/Effect Analysis (Seeking Root Cause)	
A	5. Verifying root cause with data	
	6. Solutions & Prevention steps development (including cost/benefit)	
I	7. Pilot of implementation	
	8. Implementation	
C	9. Control/Monitoring Plan (including Process Metrics)	
	10. Lessons Learned	

FIGURE 11.10
Root cause analysis.

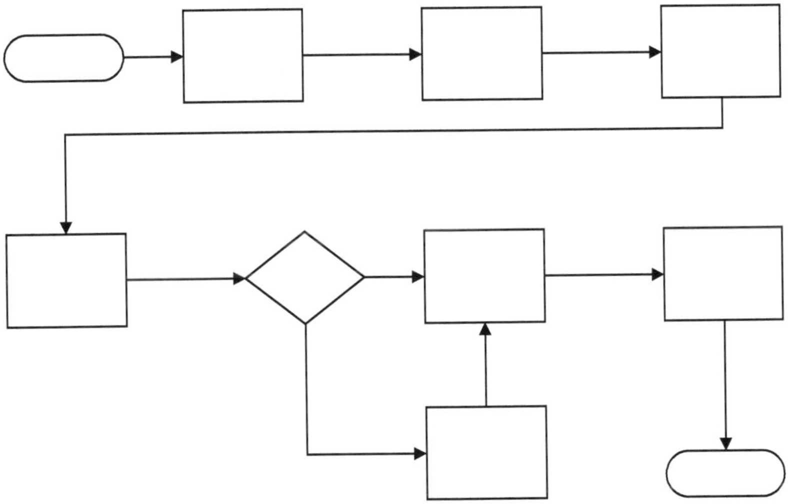

FIGURE 11.11
Process mapping.

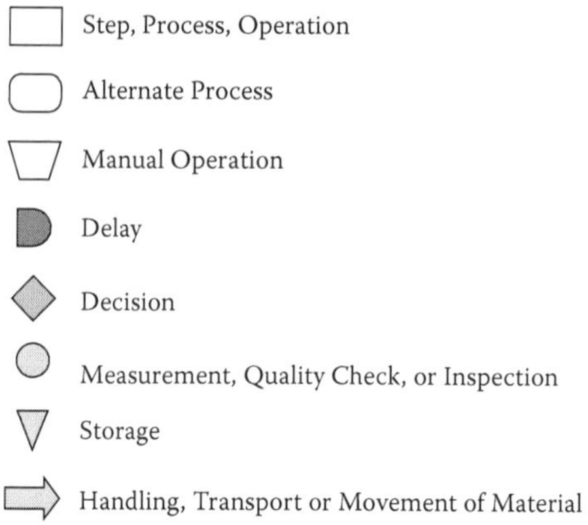

FIGURE 11.12
Process mapping symbols.

Process Mapping

Process maps are graphical representations of steps to a process. The visualization eases the complexity of the process and identifies non-value-added tasks and any key takeaways, such as redundancy and excess inspections. Process maps identify key process input variables known as *x*'s and key process output variables known as *y*'s. Any delays are to be eliminated and decisions are meant to be as efficient as possible. Process maps should be conducted for a specific area at a time due to the complexity of the processes. This way the process maps can be updated frequently by area.

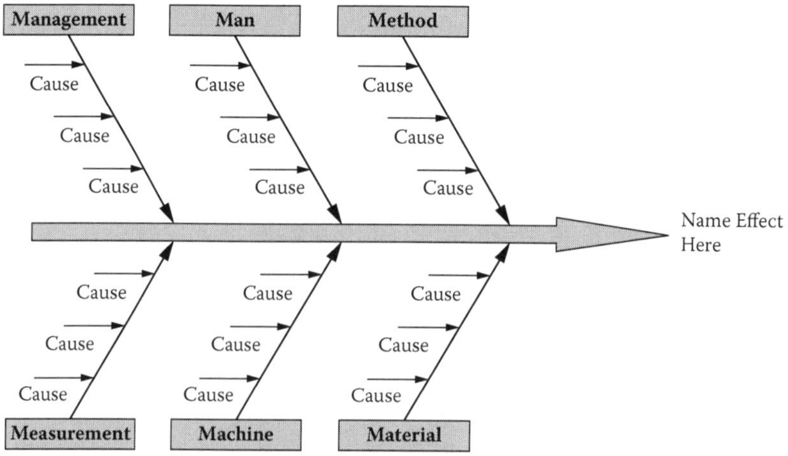

FIGURE 11.13
Fishbone diagram.

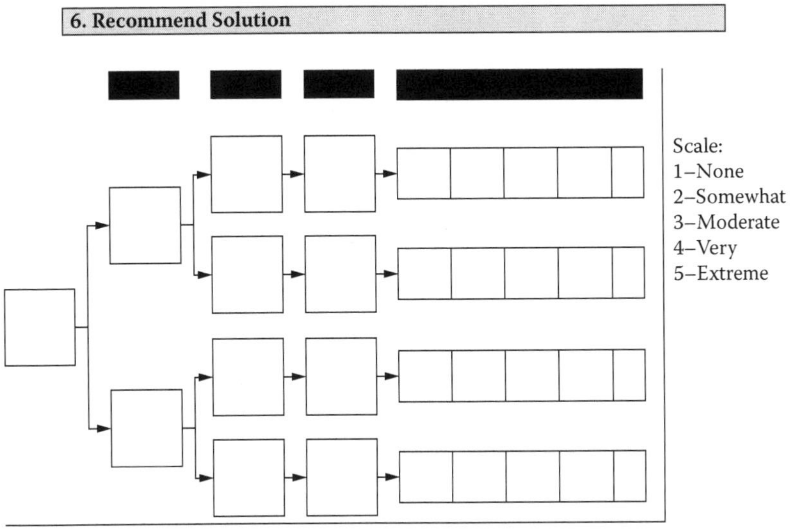

FIGURE 11.14
Recommend solutions.

Benefits and Plans

Cost Benefit		
Annualized cost of problem	1	$0.00
Percent of problem reduction	2	0%
Cost of proposed solution	3	$0.00
Total first year savings (1 x 2 - 3)		$0.00

7-8. Action plan for implementation

Who, what, when, where, and how

Root Cause Resolution

9. Control Plan (include process metrics)

10. Lessons Learned

FIGURE 11.15
Benefits and plans with action plans and lessons learned.

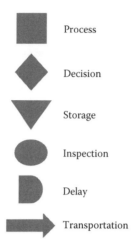

FIGURE 11.16
Process map symbols.

The process mapping symbols are as follows in Figure 11.16. An example of a process map is shown in Figure 11.17.

Financial Justification

Financial justification is important in order to be able to purchase products, utilize resources, or give return on investments. Cost versus benefits are able to be understood through financial justification. Utilizing these cost benefit tools is essential to gain approval from upper management to purchase parts or equipment, or put forth resources for a project. Normally, a return on investment of less than one year is applicable for the purchase of goods. The identification of costs along with controlling the costs is a financial attribute that should be covered. It is important to learn whether goods depreciate or

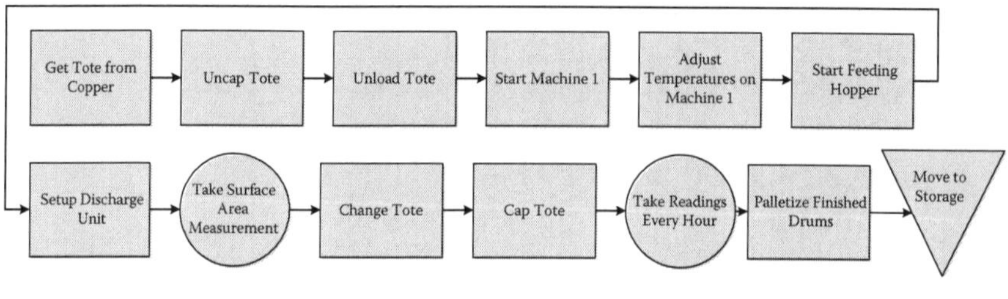

FIGURE 11.17
Process map example.

Cost Benefit		
Annualized cost of problem	1	$100,000
Percent of problem reduction	2	25%
Cost of proposed solution	3	$30,000
Total first year savings (1 × 2 – 3)		–$5,000

FIGURE 11.18
Financial justification example.

increase in value over time. Benefits should always be considered beginning with safety and ending with financial justification of a payback. Tangible results from the justification will give approval for the resources or purchase of the goods. Taxes, discounts, and depreciation should be looked at to complete financial justifications. Compounding is equivalent to adding the interest to the principal sum year on year, whereas discounting is equivalent to taking out the interest, backward in time. A cost justification example is shown in Figure 11.18, where first-year savings were not met. In this case, the payback period should be calculated. If the payback is over 5 years, normally the project is not justified.

One Point Lessons

A one point lesson is a simple, direct way to provide directions that can be easily found again and again. Visual documentation helps for this lesson so that it is not forgotten. This educational documentation ensures that employees develop their knowledge and skill sets while helping with problem-solving analyses. A one point lesson consists of the general information, which includes a title and objective. The lesson explains the description of how something is performed utilizing key points and visuals. Effective one point lessons will answer the 5W's and 1H: who, what, where, why, when, and how. An example of a one point lesson is shown in Figure 11.19.

Value Stream Mapping

Value stream mapping is similar to process mapping, although it is more advanced because it focuses on the processes utilizing Lean principles according to value. It is important to ensure that a process is considered important if the step adds value. Adding value is also defined as any activity that the customer is willing to pay for. Another note to remember is that in addition to having a smart and efficient technique, only goods that the customer is demanding should be produced to eliminate excess inventory.

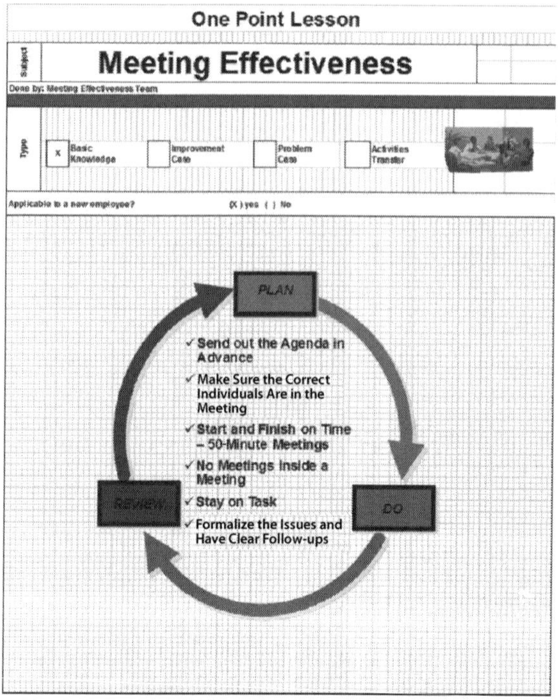
One Point Lesson
Subject
Meeting Effectiveness
Type
Basic Knowledge
Improvement Case
Problem Case
Activities Transfer
PLAN
✓ Send out the Agenda in Advance
✓ Make Sure the Correct Individuals Are in the Meeting
✓ Start and Finish on Time – 50-Minute Meetings
✓ No Meetings Inside a Meeting
✓ Stay on Task
✓ Formalize the Issues and Have Clear Follow-ups
REVIEW
DO

FIGURE 11.19
One point lesson example.

Value stream mapping reveals why excessive waste is introduced. The steps for value stream mapping are:

- Map current state processes
- Summarize current state processes
- Map future state processes
- Summarize future state processes
- Develop short-term plans to move from current to future state
- Develop long-term plans to move from current to future state
- Implement risks, failures, and processes for transitions
- Map key project owners, key dates, and future dates of future states
- Continue to map new future states when future states are met

All steps in value stream mapping should deliver a product or service to customers.

A value stream involves multiple activities that convert inputs into outputs. All processes follow sequences of starts, lead times, and ends. The first goal

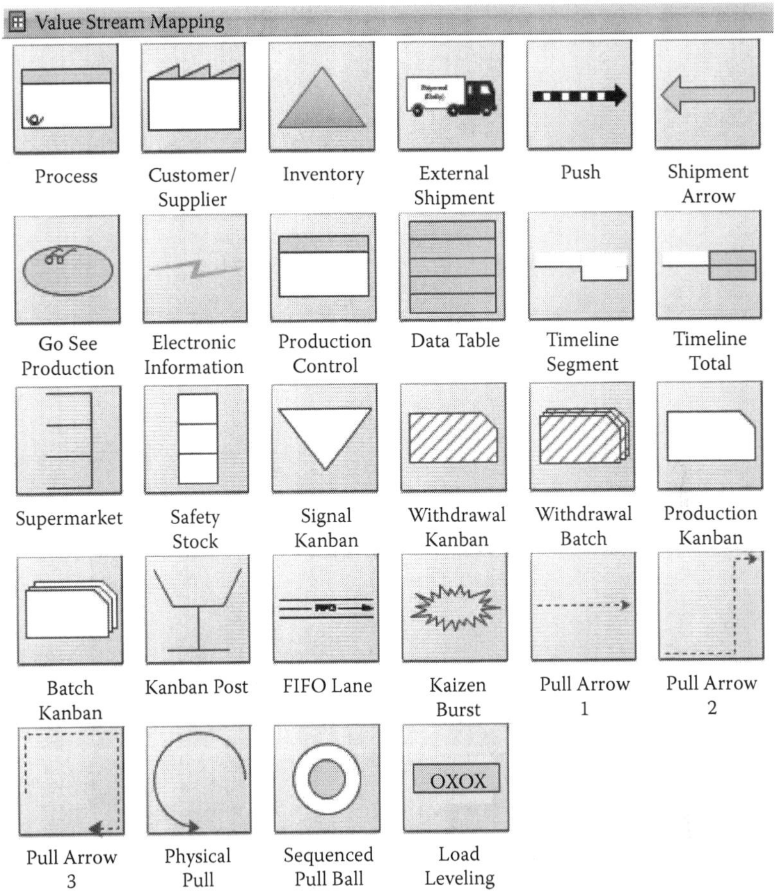

FIGURE 11.20
Value stream mapping basic symbols.

is to reduce problematic areas by understanding the main causes for defects. The next step is to reduce lead time. The final step is to find the percentage of value-added time as shown in Figure 11.20. The formula is %VAT = (Sum of activity times/Lead time) × 100. When the sum of activity times equals lead time, the value-added time is 100%. For most processes, %VAT = 5% to 25%. If the sum of activity times equals the lead time, the time value is not acceptable and activity times should be reduced. Lead time is the time from process start to end or time from receipt to delivery. Activity time is the time required to complete one output in an activity. Cycle time is the average time in which an activity yields an output, calculated as the activity time divided by the number of outputs produced in one cycle of an activity.

The value stream mapping symbols are shown in Figure 11.20. An example of an actual value stream map is shown in Figure 11.21.

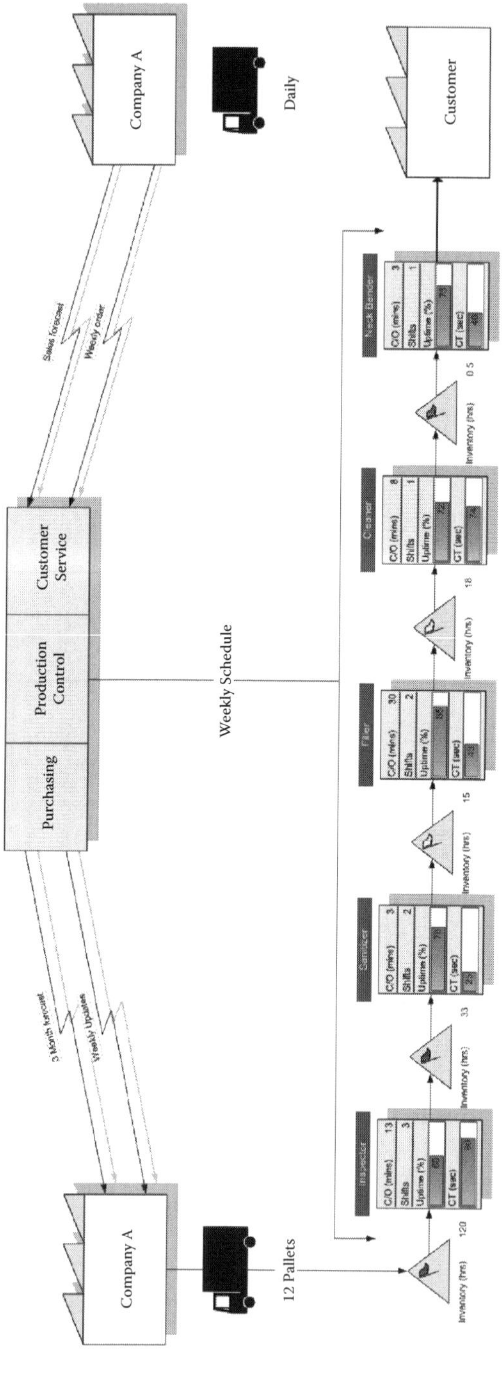

FIGURE 11.21
Value stream map example.

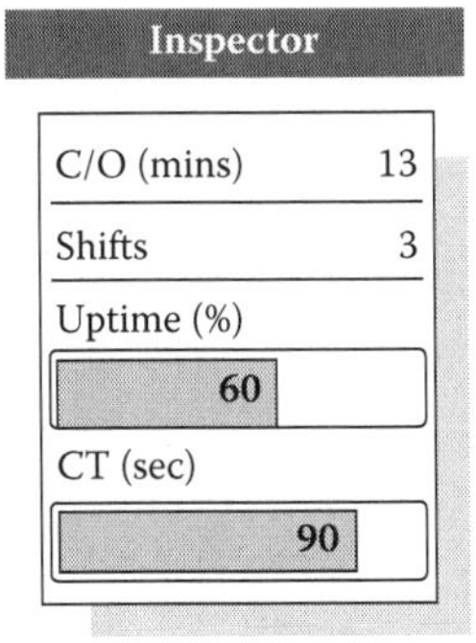

FIGURE 11.22
Value stream map cycle time.

The main takeaways from the value stream map example are the following: C/O (changeover time), Shifts, and Uptime. The cycle time can be calculated with the information showing the value-added activities (Figure 11.22).

Plan–Do–Check–Act

Plan–Do–Check–Act is a traditional cycle where processes and conditions are planned out, the planned actions are performed in the Do phase, and finally quality control checks are performed in the Check phase (Figure 11.23). This method catches mistakes and also provides feedback during the Check phase. The checks in this place also account for 100% inspection, therefore all parts or processes are looked upon indicating no defects.

There are three main types of checks or inspections that are popular:

- Judgment inspections
- Informative inspections
- Source inspections

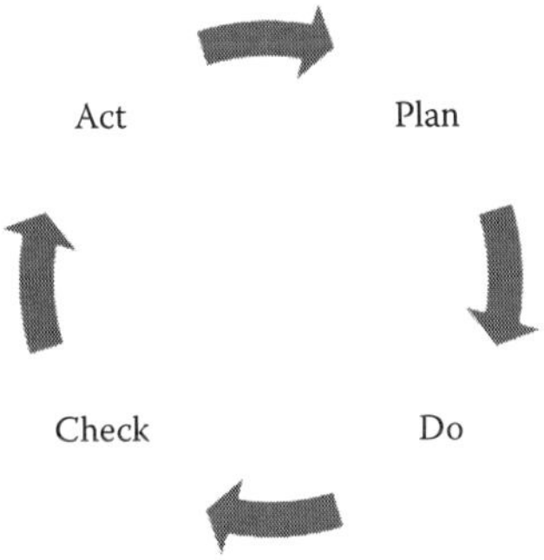

FIGURE 11.23
Plan–Do–Check–Act illustration.

Judgment inspections are those that are normally done by humans based on their expectations. They find the defect after it has already occurred. Informative inspections are based on statistical quality control (SQC) checks on each product and self-checks. These inspections help reduce defects, but do not eliminate them completely. Finally, source inspections reduce the defects completely. Source inspections discover the mistakes prior to processing and then provide feedback and corrective actions so that there are zero defects during the process. Source inspections require 100% inspection. The feedback loop is also very quick so that there is minimal waiting time.

Poka Yokes

A *poka yoke* is a Japanese term that means "mistake proofing." All inadvertent defects can be prevented from happening or prevented from being passed along. Poka Yokes use two approaches:

- Control systems
- Warning systems

Control systems stop the equipment when a defect or unexpected event occurs. This prevents the next step in the process to occur so that the complete process is not performed. Warning systems signal operators to stop the process or address the issue at the time. Obviously, the first of the two prevents all defects and has a more zero quality control (ZQC) methodology because an operator could be distracted or not have time to address the problem. Control systems often also use lights or sounds to call attention to the problem, that way the feedback loop is again very minimal.

The methods for using Poka Yoke systems are as follows:

- Contact methods
- Fixed-value methods
- Motion-step methods

Contact methods are simple methods that detect whether products are making physical or energy contact with a sensing device. Some of these are commonly known as limit switches, where the switches are connected to cylinders and pressed in when the product is in place. If a screw is left out, the product does not release to the next process. Other examples of contact methods are guide pins.

Fixed-value methods are normally associated with a particular number of parts to be attached to a product or a fixed number of repeated operations occurring at a particular process. Fixed-value methods utilize devices such as counting mechanisms. The fixed-value methods may also use limit switches or different types of measurement techniques.

Finally, the motion-step method senses if a motion or step in the process has occurred in a particular amount of time. It also detects sequencing by utilizing tools such as photoelectric switches, timers, or barcode readers.

The conclusion of Poka Yokes is to use the methodology as mistake proofing for ZQC to eliminate all defects, not just some. The types of Poka Yokes do not have to be complex or expensive, just well thought out to prevent human mistakes or accidents.

The Poka Yoke discussion leads to a correct location. This technique places design and production operations in correct order to satisfy customer demand. The concept is to increase the throughput of machines ensuring that production is performed at the proper time and place. Centralization of areas helps final assemblers, but the most common practice to be effective is to unearth an effective flow. U-shaped flows normally prevent bottlenecks. Value stream mapping is a key component during this time in order to establish that all the steps that are occurring are adding value. Another note to remember is that in addition to having a smart and efficient technique, only goods that the customer is demanding should be produced to eliminate excess inventory (Figure 11.24).

The advantages of mistake proofing include:

- No formal training required
- Relieves operators from repetitive tasks
- Promotes creativity and value-added activities
- 100% inspection internal to the operation
- Immediate action when problems arise
- Eliminates the need for many external inspection operations
- Results in defect-free work

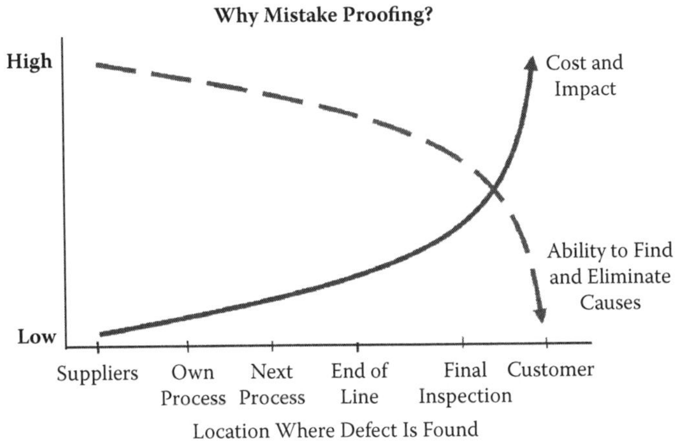

FIGURE 11.24
Why mistake proofing?

Error		Defect
Not setting the timer properly on your toaster		Burnt toast
Placing the original in your copier face up		Blank pages
Running out of link on your date stamp		No date stamp on paper

FIGURE 11.25
Mistake proofing examples.

An error occurs when the conditions for successful processing are either incorrect or absent. Defects are the result (Figure 11.25). Figure 11.26 demonstrates the use of asking why five times combined with asking why the defect escaped the process five times. Defects are prevented if:

- Errors are prevented from happening
- Errors are discovered and eliminated

How to mistake proof:

- Adapt the right attitude
- Select a process to mistake proof
- Select a defect to eliminate
- Determine the source of the defect
- Identify countermeasures
- Develop multiple solutions
- Implement the best solution
- Document the solution

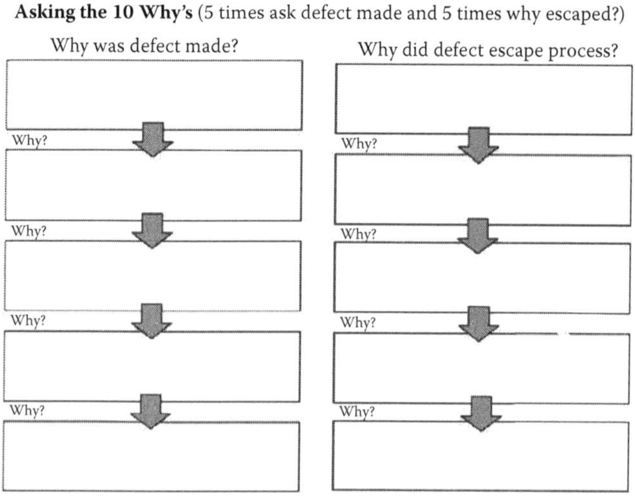

FIGURE 11.26
Ten whys for mistake proofing.

Which processes should be mistake proofed?

- High-error potential
- Complex processes
- Routine "boring" processes
- High-failure history
- Critical process characteristics
- High scores in FMEA
- Use problem history
- Pareto charts, etc.
- Isolate the specific defect
- Do not be too general
- Do not combine defects
- Make a decision on which defect to work on based on
  - Severity
  - Frequency
  - Ease of solving
- Annoyance factor

Figure 11.27 explains the successive checking positives and negatives.

Successive Checks	Self-Check Detection in Station	Mistake Proof (Poka Yoke)
Associates check work of previous associate	Associates check own work before passing to the next associate	Automatic check and prevention of defect
Plus: Generally effective in catching defects	Plus: Instant correction possible and more palatable than supervisor check or peer check	Plus: 100% inspection usually with no extra time expense with the benefit of instant correction
Corrective action can only occur after defect is made	Associate may compromise quality or forget to perform self-check	None

FIGURE 11.27
Different methods of checks.

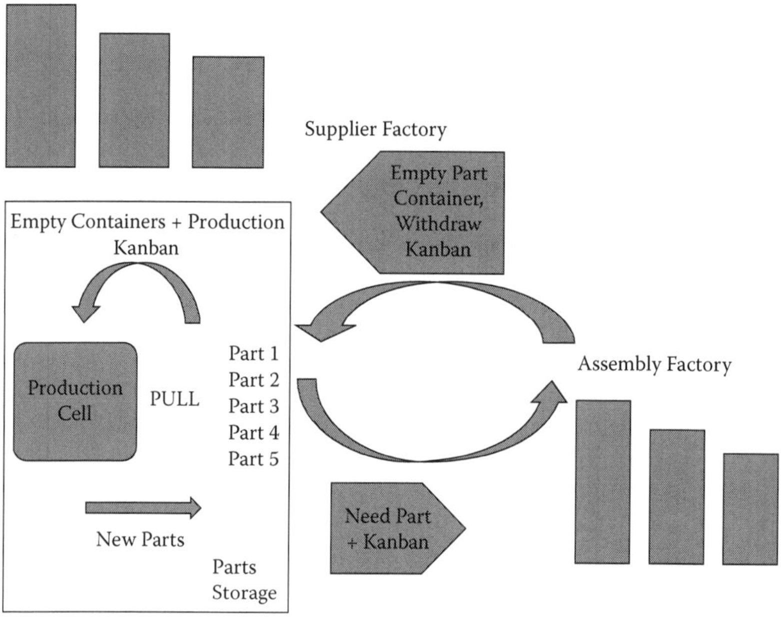

FIGURE 11.28
Kanban example.

Kanbans

Kanbans are communication signals that control inventory levels while ensuring even and controlled production flow. Kanbans signal the times to start, times to change setups, and times to supply parts. Kanbans work only if consistently monitored. Kanbans are used in Lean production to ensure that flow is pulled by the next step and a visual indicator signals the need for inventory or activity (Figure 11.28). These steps enable production to respond directly to customer needs without producing excess inventory or requiring further work. Quality should also soar because production is based on customer needs and not production per minute or per hour.

Pull and Push Flows

Pull is the simple practice of not producing a product upstream until the customer downstream needs it. The philosophy sounds easy, but its management is complicated. Normally, large batches of general inventory are in stock, but ensuring specialized and costly inventories for specific customers is far more difficult. Understanding how to be versatile and utilizing co-manufacturers for these special circumstances are keys to Lean production utilizing a pull system. Customers for high-end or specialized products must understand the lead times required.

Push systems are not accurate because they normally involve production schedules based on projected customer demands. Changeovers should be kept to a minimum to produce versatile products instead of large quantities of a special product. Small "stores" of parts between operations are created to reduce inventory. When a process customer uses inventory, it should be replenished. If inventory is not needed, it is not replenished. Lean production for pull requires 90% machine use or availability, and limits downtime for maintenance and changeovers to 10% of machine time.

Visual Management

Visual boards show team successes and provide communication without the need for every team member to be present. A visual management system normally consists of some type of large whiteboard that has key points, which are important to each team member. Common visual boards consist of the following points:

- Each person writes down three key items that they are working on and what their holdup is
- Recognition of a key associate
- Capital spending box
- Reminders
- Calendar reference
- Meeting time
- Overdue action items

It is important to meet once a week at the same time to review the visual management board. Each person is accountable for their box and should fill it out prior to the meeting.

Utilizing different colors for certain topics, such as pink for important items, helps people to visually separate items. Suggestions from the visual management boards should be used, because the board should always be

FIGURE 11.29
Visual management technique.

changing and improving. It is also important to remember that people react to colors, shapes, and so forth, so these should be utilized (Figure 11.29).

Any types of measurements that can be seen visually are also key indicators of whether or not a process is performing well. A visual technique for this is shown in a simple measuring cup in Figure 11.30.

FIGURE 11.30
Visual management technique.

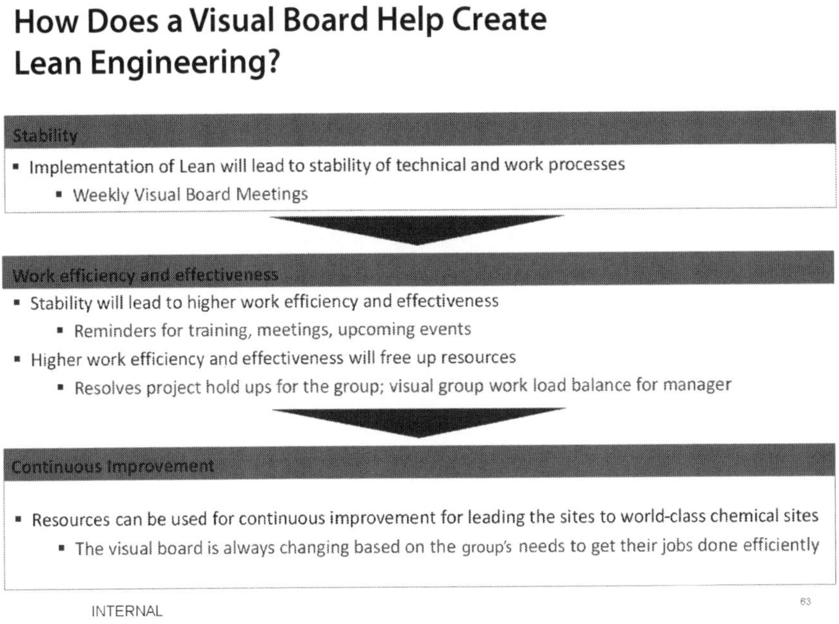
How Does a Visual Board Help Create Lean Engineering?
Stability
▪ Implementation of Lean will lead to stability of technical and work processes
▪ Weekly Visual Board Meetings
Work efficiency and effectiveness
▪ Stability will lead to higher work efficiency and effectiveness
▪ Reminders for training, meetings, upcoming events
▪ Higher work efficiency and effectiveness will free up resources
▪ Resolves project hold ups for the group; visual group work load balance for manager
Continuous Improvement
▪ Resources can be used for continuous improvement for leading the sites to world-class chemical sites
▪ The visual board is always changing based on the group's needs to get their jobs done efficiently
INTERNAL
63

FIGURE 11.31
Visual management board and Lean engineering.

The purpose of these visual controls is to show what's right, what's wrong, what's complete, where the path going forward should be, what step to take next, who to see, and common standard operating techniques (Figure 11.31). These techniques tell us at a glance how things are going, and instantaneously points out imperfections or abnormal conditions. Visual techniques will enhance performance to the next level improving office and factory performances.

Cellular Processing

Centrally locating cells within all machines and workstations is important so that the final product can be completed without major transportation delays. The premise is to reduce inventory, lead time, material handling, intermediate storage, and improve communication between operations. One operator should be able to run multiple processes by utilizing centralized cellular management.

Processes and equipment that are alike should be placed together. Takt times can be calculated to increase uptime and standardize process times. Utilizing a U-shaped layout ensures that personnel is centrally located with ease of access to machinery and workstations. The quality of inspection is increased utilizing this layout as well. The capacity is increased in this methodology by only making for customer orders by calculating the demand needed for all items. Cycle times should be calculated for all parts of machinery and processing times, operator load times, and setup and changeover times. If the capacity of one operation is higher than the demand for the product, the machinery should be able to sit idle. The bottleneck in this case should be identified, and the downtime should be eliminated for it. The reduction of setup times and changeover times is the ultimate goal.

There are five main steps to make this cellular layout:

- Group similar products
- Measure demands through Takt time
- Review work sequences
- Combine work in balanced processes
- Design the cellular layout

U-shaped layouts are generally the most efficient since they provide the shortest distance for time traveled. Work also enters and exits the same vicinity while communication is increased and flexibility can be completed by different operators who are able to be cross-functional. A modified circle is also another option that has the same premises of the U shape.

Spaghetti Diagrams

To map the current flow to reduce time, an analysis of movement should be performed. Simple ways to map flow is by utilizing a spaghetti diagram. Flows should be continuous, and visually illustrating them is an easy way to determine travel distances. A spaghetti diagram can be completed by simply drawing out the layout of equipment (Figure 11.32) and then mapping a person's walking paths throughout the production day. Virtually all steps should be mapped with lines showing where the person travels (Figure 11.33). After the travel time is mapped, it can be seen where the bottlenecks lie. The step after this is to perform the future state of the spaghetti diagram.

The steps for mapping a spaghetti diagram are as follows:

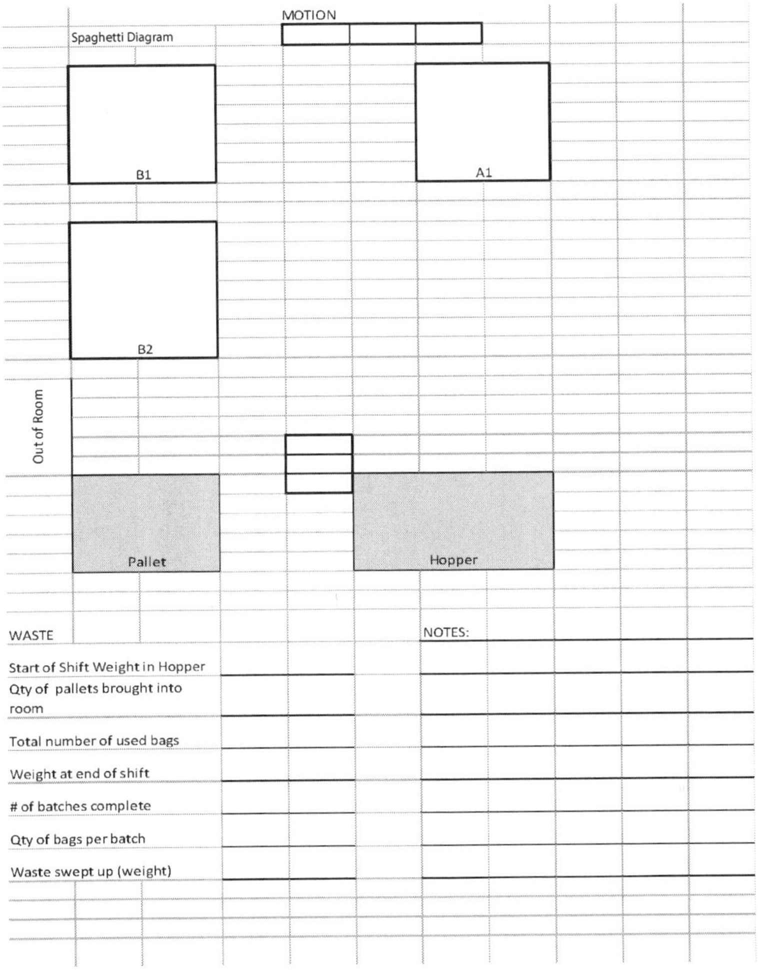

FIGURE 11.32
Spaghetti diagram layout.

- *Step 1*—Create a map of the work area layout
- *Step 2*—Observe the current work flow and draw the actual work path from the very beginning of work to the end when products exit the work area
- *Step 3*—Analyze the spaghetti chart and identify improvement opportunities

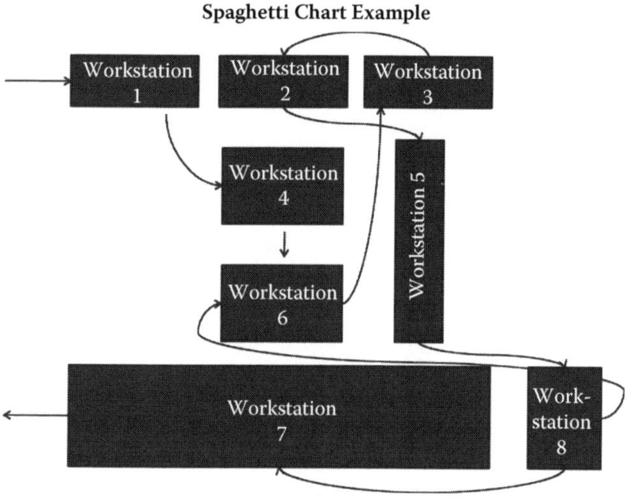

FIGURE 11.33
Spaghetti diagram.

Histograms

A histogram plots the frequency of values grouped together as a bar graph. Histograms are handy for determining location, spread, and shape. Outliers can easily be identified. The height equals the frequency and the width equals a range of values. A histogram with a bell-shaped curve is normal (Figure 11.34).

Pareto Charts

Key projects can be determined by performing Pareto analysis. This statistical tool implies that by doing 20% of the work, 80% of the advantages can be generated. When applied to quality, this philosophy states that 80% of problems stem from 20% of key causes. Pareto analyses are guides to prioritizing and determining key opportunities. See Figure 11.35 for an example of a Pareto chart.

Capability Analysis

Industrial process capability analysis is an important aspect of managing industrial projects. The capability of a process is the spread that contains most of the values of the process distribution. It is very important to note

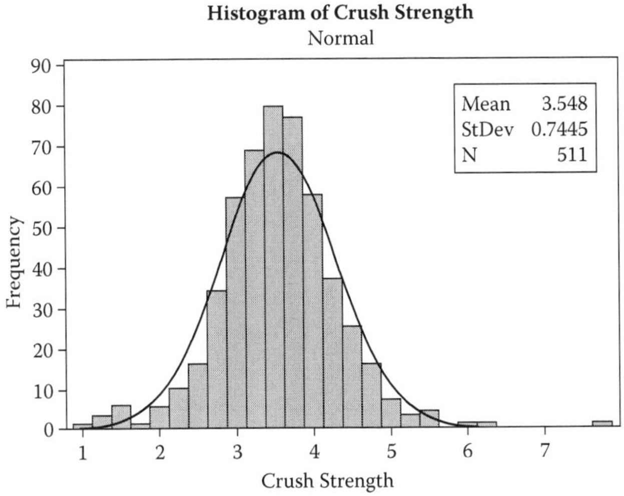

FIGURE 11.34
Histogram example.

that capability is defined in terms of distribution. Therefore, capability can only be defined for a process that is stable (has distribution) with common cause variation (inherent variability). It cannot be defined for an out-of-control process (no distribution) with variation arising from specific causes (total variability). The key need for capability analysis is to ensure the process is meeting the requirements. Capability analysis can be done with both attribute and variable data. They measure the short- and long-term process

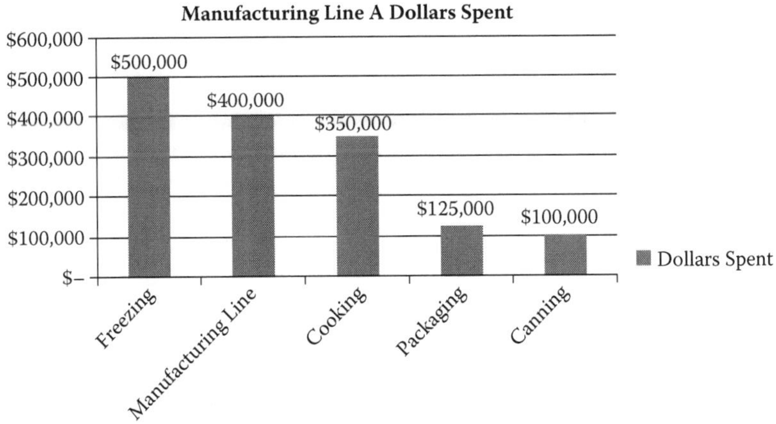

FIGURE 11.35
Pareto chart.

capabilities. The key capability indices are C_p and C_{pk}, or P_p and P_{pk} for long-term capabilities. The following formulas are used for capability indices:

$$C_p = \frac{USL - LSL}{6\hat{\sigma}}$$

$$C_pU = \frac{USL - \bar{X}}{3\hat{\sigma}}$$

$$C_pL = \frac{\bar{X} - LSL}{3\hat{\sigma}}$$

$$C_{pk} = Min(C_pU,\ C_pL)$$

The value of C_{pk} should be greater than 1.5 to be of a "good" capability. C_p and C_{pk} use a pooled estimate of the standard deviation, whereas P_p and P_{pk} use the long-term estimate of the standard deviation. C_{pk} is what the process is capable of doing if there is no subgroup variability and P_{pk} is the actual process performance. Normally, C_{pk} is smaller than P_{pk} since P_{pk} represents both within-subgroup and between-subgroup variability. C_{pk} only represents between subgroup variability. The main steps to a capability study consist of the following:

1. Set up the process to the best parameters and identify Key Process Input Variables
2. Identify subgroups
3. Run the product over a short time span to minimize the impact of special cause variation
4. Observe the process and take notes throughout
5. Measure and identify Key Process Output Variables
6. Run capability analyses to review Normality, Statistical Process Control, and Histograms
7. Run capability analyses for short-term and total standard deviations
8. Identify mean shifts and variation
9. Estimate long-term capability
10. Develop action plans based on the above data

Setting short-term and long-term goals based on capability analyses will result in successful action plans based on real-time data.

The goal of capability studies is to do the following:

- Move the P_{pk} to P_p or to center the process
- Move the P_p to C_{pk} or to reduce the variation
- Move the C_{pk} to C_p or have random variation

A Six Sigma process has a C_p of 2.00 and a P_{pk} of 1.5.

Control Charts

Control charts may be used to track project performance before deciding what control actions are needed. Control limits are incorporated into the charts to indicate when control actions should be taken. Multiple control limits may be used to determine various levels of control points. Control charts may be developed for costs, scheduling, resource utilization, performance, and other criteria. Control charts are used extensively in quality control work to identify when a system has gone out of control. The same principle is used to control the quality of work sampling studies. The 3σ limit is normally used in work sampling to set the upper and lower limits of control. First, the value of p is plotted as the center line of a p chart. The variability of p is then found for the control limits. Two of the most commonly used control charts in industry are X-bar charts and range (R-)charts. The type of chart depends on the kind of data collected: variable data or attribute data. The success of quality improvement depends on (1) the quality of data available and (2) the effectiveness of the techniques used for analyzing the data. The charts generated by both types of data are:

Variable data

- Control charts for individual data elements (X)
- Moving range chart (MR-chart)
- Average chart (X-chart)
- Range chart (R-chart)
- Median chart
- Standard deviation chart (σ-chart)
- Cumulative sum chart (CUSUM)
- Exponentially weighted moving average (EWMA)

Attribute data

- Proportion or fraction defective chart (p-chart); subgroup sample size can vary
- Percent defective chart (100p-chart); subgroup sample size can vary
- Number defective chart (np-chart); subgroup sample size is constant
- Number defective (c-chart); subgroup sample size is one

Defective per inspection unit (u-chart); subgroup sample size can vary

The statistical theory useful to generate control limits is the same for all the charts except the EWMA and CUSUM charts.

X-Bar and Range Charts

The R-chart is a time plot useful for monitoring short-term process variations. The X-bar chart monitors longer-term variations where the likelihood of special causes is greater over time. Both charts utilize control lines called upper and lower control limits and central lines; the types of lines are calculated from process measurements. They are not specification limits or percentages of the specifications or other arbitrary lines based on experience. They represent what a process is capable of doing when only common cause variation exists, in which case the data will continue to fall in a random fashion within the control limits and the process is in a state of statistical control. However, if a special cause acts on a process, one or more data points will be outside the control limits and the process will no longer be in a state of statistical control.

Calculation of Control Limits

- Range (R)—This is the difference between the highest and lowest observations

$$R = X_{highest} - X_{lowest}$$

- Center lines—Calculate $\bar{X}$ and $\bar{R}$

$$\bar{X} = \frac{\sum X_i}{m}$$

$$\bar{R} = \frac{\sum R_i}{m}$$

where
$\bar{X}$ = Overall process average
$\bar{R}$ = Average range
m = Total number of subgroups
n = Within-subgroup sample size

- Control limits based on R-chart

$$UCL_R = D_4\bar{R}$$

$$LCL_R = D_3\bar{R}$$

- Estimate of process variation

$$\hat{\sigma} = \frac{\bar{R}}{d_2}$$

- Control limits based on $\bar{X}$-chart—Calculate the upper and lower control limits for the process average

$$UCL = \bar{X} + A_2\bar{R}$$

$$LCL = \bar{X} - A_2\bar{R}$$

The values of d_2, A_2, D_3, and D_4 are for different values of n. These constants are used for developing variable control charts.

Plotting Control Charts for Range and Average Charts

- Plot the range chart (R-chart) first.
- If the R-chart is in control, then plot the X-bar chart.
- If the R-chart is not in control, identify and eliminate special causes, then delete points that are due to special causes, and recompute the control limits for the range chart. If process is in control, then plot the X-bar chart.
- Check to see if the X-bar chart is in control, if not search for special causes and permanently eliminate them.
- Remember to perform the eight trend tests.

Plotting Control Charts for Moving Range and Individual Control Charts

- Plot the moving range chart (MR-chart) first.
- If the MR-chart is in control, then plot the individual chart (X).
- If the MR-chart is not in control, identify and eliminate special causes, then delete special cause points, and recompute the control limits for the moving range chart. If the MR-chart is in control, then plot the individual chart.
- Check to see if the individual chart is in control, if not search for special causes from out-of-control points.
- Perform the eight trend tests.

Defects per Million Opportunities

Six Sigma provides tools to improve the capabilities of business processes while reducing variations. It leads to defect reduction and improved profits and quality. Six Sigma is a universal scale that compares business processes based on their limits to meet specific quality limits. The system measures defects per million opportunities (DPMOs). The Six Sigma name is based on a limit of 3.4 defects per million opportunities.

Figure 11.36 shows a normal distribution that underlies the statistical assumptions of the Six Sigma model. The Greek letter σ (sigma) marks the distance on the horizontal axis between the mean μ and the curve inflection point. The greater this distance, the greater is the spread of values encountered. The figure shows a mean of 0 and a standard deviation of 1, that is, $\mu = 0$ and $\sigma = 1$. The plot also illustrates the areas under the normal curve within different ranges around the mean. The upper and lower specification limits (USL and LSL) are ±3 σ from the mean or within a six-sigma spread. Because of the properties of the normal distribution, values lying as far away as ±6 σ from the mean are rare because most data points (99.73%) are within ±3 σ from the mean except for processes that are seriously out of control.

Six Sigma allows no more than 3.4 defects per million parts manufactured or 3.4 errors per million activities in a service operation. To appreciate the effect of Six Sigma, consider a process that is 99% perfect (10,000 defects

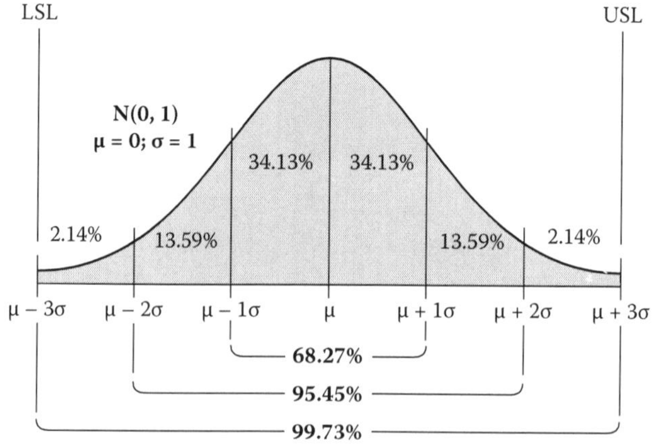

FIGURE 11.36
Areas under a normal curve.

TABLE 11.1

Defects per Million Opportunities and Sigma Levels

Sigma Spread	DPMO	Percent Defective	Percent Yield	Short-Term C_{pk}	Long -Term C_{pk}
1	691,462.00	69%	31%	0.33	–0.17
2	308,538.00	31%	69%	0.67	0.17
3	66,807.00	7%	93.3%	1	0.5
4	6,210.00	0.62%	99.38%	1.33	0.83
5	233.00	0.02%	99.98%	1.67	1.17
6	3.40	0%	100%	2	1.5

per million parts). Six Sigma requires the process to be 99.99966% perfect to produce only 3.4 defects per million, that is, 3.4/1,000,000 = 0.0000034 = 0.00034%. That means that the area under the normal curve within ±6 σ is 99.99966% with a defect area of 0.00034%. Six Sigma pushes the limit of perfection! Table 11.1 depicts long-term DPMO values that correspond to short-term sigma levels.

Project Charters

A project charter is a definition of the project that includes the following:

- Provides Problem Statement
- Overview of Scope, Participants, Goals, and Requirements
- Provides authorization of a new project
- Identifies roles and responsibilities

Once the project charter is approved, it should not be changed. A project charter begins with the project name, the department of focus, the focus area, and the product or process. An example of the table of contents for a project charter is shown in Figure 11.37.

A project charter serves as the focus point throughout the project to ensure the project is on track and the proper people are participating and being held accountable.

The importance of a project charter in relation to sustainability is to have a living document to educate and provide governance for a new project. Sustainability needs to utilize a great deal of education while giving goals

TABLE OF CONTENTS		
1	**PROJECT CHAPTER PURPOSE**	3
2	**PROJECT EXECUTIVE SUMMARY**	3
3	**PROJECT OVERVIEW**	4
4	**PROJECT SCOPE**	4
	4.1 Goals and Objectives	4
	4.2 Departmental Statements of Work (SOW)	5
	4.3 Organizational Impacts	5
	4.4 Project Deliverables	5
	4.5 Deliverables Out of Scope	5
	4.6 Project Estimated Costs $ Duration	5
5	**PROJECT CONDITIONS**	6
	5.1 Project Assumptions	6
	5.2 Project Issues	6
	5.3 Project Risks	6
	5.4 Project Constraints	6
6	**PROJECT STRUCTURE APPROACH**	6
7	**PROJECT TEAM ORGANIZATION PLANS**	7
8	**PROJECT REFERENCES**	8
9	**APPROVALS**	8
10	**APPENDICES**	9
	10.1 Document Guidelines	9
	10.2 Project Charter Document Sections Omitted	9

FIGURE 11.37
Project charter example.

and objectives. A project charter will serve as this living document for organizations with specified approaches.

SIPOC

The SIPOC identifies:

1. Major tasks and activities
2. The boundaries of the process
3. The process outputs
4. Who receives the outputs (the customers)
5. What the customer requires of the outputs
6. The process inputs
7. Who supplies the inputs (suppliers)

Supplier	Input	Process	Output	Customer

FIGURE 11.38
SIPOC.

8. What the process requires of the inputs
9. The best metrics to measure

SIPOC stands for the following (also see Figure 11.38):

Supplier—Know and work with your supplier while ensuring that your supplier improves their processes.

Input—Strive to continually improve the inputs by trying to do the right thing the first time.

Process—Describe the process at a high level, but with enough detail to demonstrate to an executive or manager. Understand the process fully by knowing it 100%. Eliminate any mistakes by doing a Poka Yoke.

Output—Strive to continually improve the outputs by utilizing metrics.

Customer—Keep the customer's requirements in sight by remembering that they are the most important aspect of the project. The customer makes the specifications; keep the CTQs (critical to qualities) of the customer in mind.

SIPOC steps:

1. Gain top-level view of the process
2. Identify the process in simple terms
3. Identify External Inputs such as raw materials, employees, and so on
4. Identify the Customer Requirements, also known as Outputs
5. Make sure to include all value-added and non-value-added steps
6. Include both process and product output variables

SIPOC implies that the process is understood and helps easily identify opportunities for improvement.

A SIPOC is important in concepts of sustainability because it helps develop a solution for development. Normally, the process is mapped out in a well-defined manner but in a high-level philosophy.

Figure 11.39 is an example of a SIPOC for assembling a part.

The important part of a SIPOC is to look at the details of the current state and see what improvements can be made for future states. Adding specifications for any of the inputs can identify gaps in the process. Benchmarking one process to another will also identify gaps.

Suppliers	Inputs	Process	Outputs	Customers
Operations	Manufacturing Date	Assemble Part A	Finished Part A	Customer
Sales	Prices	Tighten Part A	Receipt	Sales
Finance	Pay Back	Calibrate Part A	Invoice	Accounting
		QC Part A	Packing Slip	
		Pack Part A		

FIGURE 11.39
SIPOC.

Kano Model

The Kano Model was developed by Noriaki Kano in the 1980s. The Kano Model is a graphical tool that further categorizes VOC (Voice of the Customer) and CTQs into three distinct groups:

- Must Haves
- Performance
- Delighters

The Kano helps identify CTQs that add incremental value versus those that are simply requirements where having more is not necessarily better.

The Kano Model engages customers by understanding the product attributes that are most likely important to customers. The purpose of the tool is to support product specifications that are made by the customer and promote discussion while engaging team members. The model differentiates features of products rather than customer needs by understanding necessities and items that are not required whatsoever. Kano also produced a methodology for mapping consumer responses with questionnaires that focused on attractive qualities through reverse qualities. The five categories for customer preferences are as follows:

- Attractive
- One-dimensional

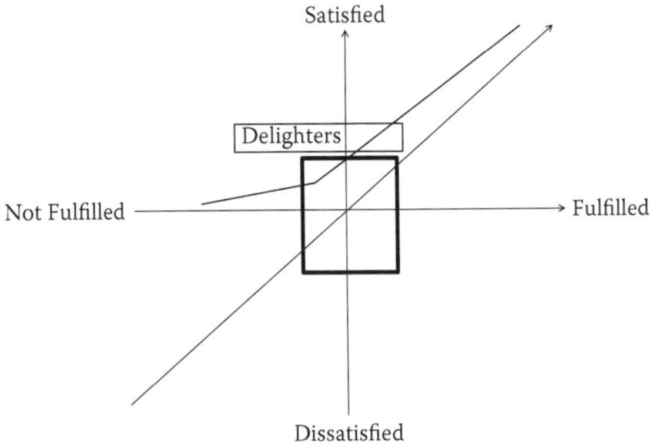

FIGURE 11.40
Kano model.

- Must-be
- Indifferent
- Reverse

Attractive qualities are those that provide satisfaction when fulfilled, but do not result in dissatisfaction if not fulfilled. One-dimensional qualities are those that provide satisfaction when fulfilled, and dissatisfaction if not fulfilled. Must-be qualities are those that are taken for granted if fulfilled, but provide dissatisfaction when not fulfilled. Indifferent qualities are those that are neither good nor bad resulting in neither customer satisfaction nor dissatisfaction. Reverse qualities are those that result in high levels of dissatisfaction from some customers and show that most customers are not alike (Figure 11.40). A Kano table is shown in Table 11.2. The Kano Model is important to use when being sustainable because it is important to differentiate which aspects we must accomplish to protect our environment and which aspects we can gradually improve upon.

TABLE 11.2
Kano Model Table

	Answers to a negatively formulated question				
		I like that	That's normal	I don't care	I don't like that
	I like that		Delighter	Delighter	Satisfier
	That's normal				Dissatisfier
	I don't care				Dissatisfier
Answers to a positively formulated question	I don't like that				

Critical to Quality (CTQ)

CTQ or Critical to Quality are characteristics that are important to the customer. They are measureable and quantifiable metrics that come from the Voice of the Customer (VOC). An affinity diagram is an organizational tool for VOCs.

Critical to Quality is important to communication because we need to understand the aspects that matter most to the customer.

Utilizing a VOC for manufacturing internally is a good way to understand processes that the employees know a great deal about. Therefore, the production worker(s) are the customers and the questions are given to them.

A tree format helps with the visualization of CTQs (Figure 11.41).

Affinity Diagram

An affinity diagram is a tool conducted to place large amounts of information into an organized manner by grouping the data into characteristics (Figure 11.42). The steps for an affinity diagram are as follows:

- *Step 1*—Clearly define the question or focus of the exercise
- *Step 2*—Record all participant responses on notecards or sticky notes
- *Step 3*—Lay out all notecards or post the sticky notes onto a wall
- *Step 4*—Look for and identify general themes
- *Step 5*—Begin moving the notecards or sticky notes into the themes until all responses are allocated
- *Step 6*—Reevaluate and make adjustments

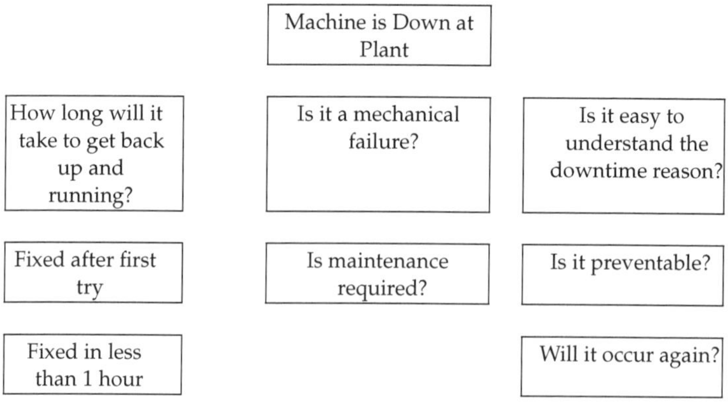

FIGURE 11.41
CTQ tree.

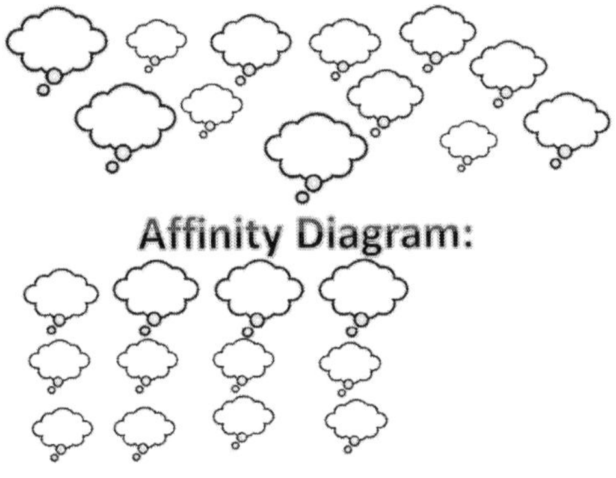

FIGURE 11.42
Affinity diagram example.

The steps to an affinity diagram for a basic process are:

- Record the date of the process or problem before the affinity diagram is completed on a piece of paper or sticky note
- Move the notes around so that there are clusters of similar themes in a group (this is normally done by a group)
- Do not explain the thoughts behind the movements of the notes
- Continue the movement of the notes into clusters until everyone seems pleased
- Once this process is completed, a discussion is to be completed on the reason the groups are as they are
- Problematic reasons can be more visible with the clusters
- Problem solving can be completed from the clusters

The pros and cons are then sought after to have a decision. The decision should be made by having a consensus from the group where the pros outweigh the cons.

Measurement Systems Analysis

Gage R&R

Gage R&R is a measurement systems analysis (MSA) technique that uses continuous data based on the principles that:

- Data must be in statistical control.
- Variability must be small compared to product specifications.

- Discrimination should be about one-tenth of product specifications or process variations.
- Possible sources of process variation are revealed by measurement systems.
- Repeatability and reproducibility are primary contributors to measurement errors.
- The total variation is equal to the real product variation plus the variation due to the measurement system.
- The measurement system variation is equal to the variation due to repeatability plus the variation due to reproducibility.
- Total (observed) variability is an additive of product (actual) variability and measurement variability.

Discrimination is the number of decimal places that can be measured by the system. Increments of measure should be about one-tenth of the width of a product specification or process variation that provides distinct categories.

Accuracy is the average quality near to the true value.

The *true value* is the theoretically correct value.

Bias is the distance between the average value of the measurement and the true value, the amount by which the measurement instrument is consistently off target, or systematic error. *Instrument accuracy* is the difference between the observed average value of measurements and the true value. Bias can be measured based by instruments or operators. Operator bias occurs when different operators calculate different detectable averages for the same measurement. Instrument bias results when different instruments calculate different detectable averages for the same measurement.

Precision encompasses total variation in a measurement system, the measure of natural variation of repeated measurements, and repeatability and reproducibility.

Repeatability is the inherent variability of a measurement device. It occurs when repeated measurements are made of the same variable under absolutely identical conditions (same operators, setups, test units, environmental conditions) in the short term. Repeatability is estimated by the pooled standard deviation of the distribution of repeated measurements and is always less than the total variation of the system.

Reproducibility is the variation that results when measurements are made under different conditions. The different conditions may be operators, setups, test units, or environmental conditions in the long term. Reproducibility is estimated by the standard deviation of the average of measurements from different measurement conditions.

The *measurement capability index* is also known as the precision-to-tolerance (P/T) ratio. The equation is $P/T = (5.15 \times \sigma MS)/\text{tolerance}$. The P/T ratio is usually expressed as a percent and indicates what percent of the tolerance is taken up by the measurement error. It considers both repeatability and the reproducibility. The ideal ratio is 8% or less; an acceptable ratio is 30% or less. The 5.15 standard deviation accounts for 99% of the measurement system (MS) variation and is an industry standard.

The P/T ratio is the most common estimate of measurement system precision. It is useful for determining how well a measurement system can perform with respect to the specifications. The specifications, however, may be inaccurate or need adjustment. The %R&R = (σMS/σTotal) × 100 formula addresses the percent of the total variation taken up by measurement error and includes both repeatability and reproducibility.

A Gage R&R can also be performed for discrete data, also known as binary data. This data is also known as yes/no or defective/nondefective type data. The data still requires at least 30 data points. The percentages of repeatability, reproducibility, and compliance should be measured. If no repeatability is able to be shown, there will also be no reproducibility. The matches should be above 90% for the evaluations. A good measurement system will have a 100% match for repeatability, reproducibility, and compliance.

If the result is below 90%, the operational definition must be revisited and redefined. Coaching, teaching, mentoring, and standard operating procedures should be reviewed, and the noise should be eliminated.

A Gage R&R is shown in Figure 11.43 where there is a decision to be made on what equipment is sustainable and what employees are sustainable in a factory based on measurement data.

The Gage R&R bars are desired to be as small as possible, driving the Part-to-Part bars to be larger.

The averages of each operator is different, meaning the reproducibility is suspect. The operator is having a problem making consistent measurements.

The Operator*Samples Interactions lines should be reasonably parallel to each other. The operators are not consistent to each other.

The Measurement by Samples graph shows there is minimal spread for each sample, and a small amount of shifting between samples.

The Measurement by Operators shows that the operators are not consistent, and Operator 2 is normally lower than the rest.

In Table 11.3, the Sample times Operator of .706 shows that the interaction was not significant, which is what is wanted from this study.

The %Contribution Part to Part of 10.81 shows the parts are the same.

The Total Gage R&R % Study Variation of 94.44, %Contribution of 89.19, Tolerance of 143.25, and Distinct Categories of 1 showed this there was no repeatability, reproducibility, and was not a good gage. The number of categories being less than 2 shows the measurement system is of minimal value since it will be difficult to distinguish one part from another.

The gage is a bad representation based on the results shown in Figure 11.44. The Gage Run Chart (Figure 11.45) shows that there is no consistency between measurements.

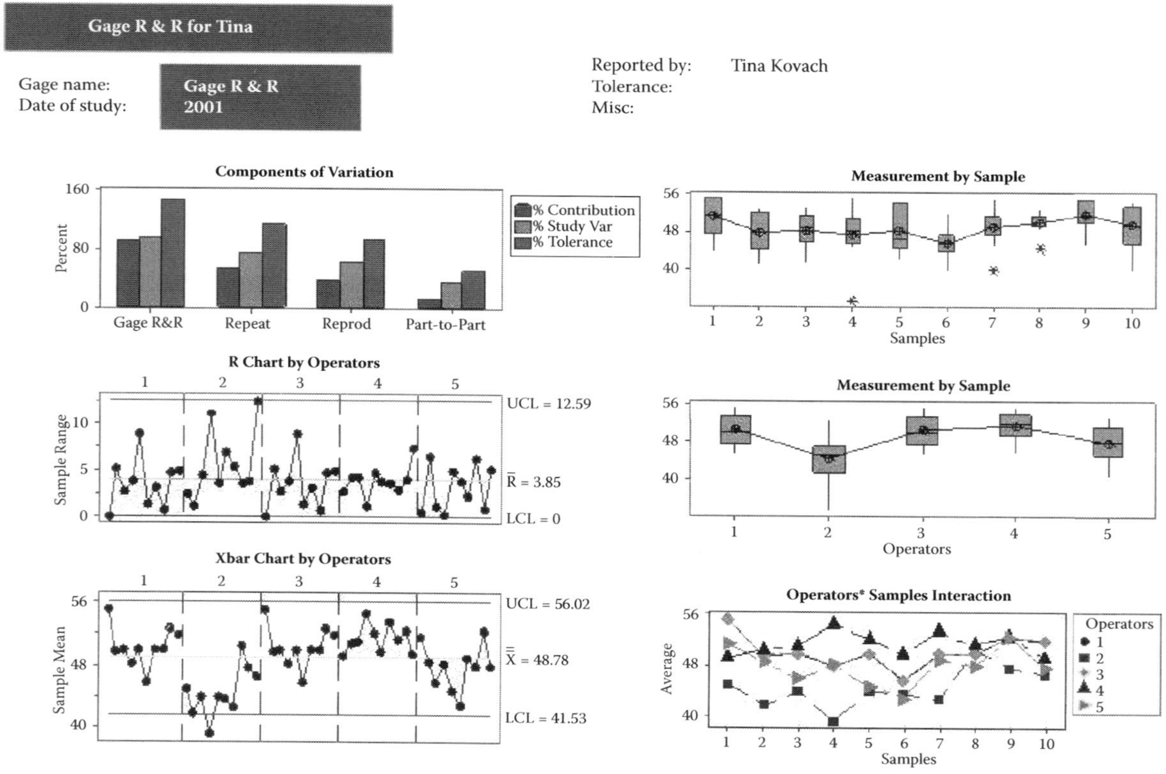

FIGURE 11.43
Gage R&R example.

TABLE 11.3

Gage R&R Results

Gage R&R Study—ANOVA Method

Gage R&R for Measurement

Gage Name: White Butter Creme Gage R&R

Date of study: 11/18/10

Reported by: Tina Kovach

Tolerance:

Misc:

Two-Way ANOVA Table with Interaction

Source	DF	SS	MS	F	P
Samples	9	282.49	31.388	3.3908	0.004
Operators	4	611.14	152.785	16.5050	0.000
Samples * Operators	36	333.25	9.257	0.8398	0.706
Repeatability	50	551.13	11.023		
Total	99	1778.01			

Alpha to remove interaction term = 0.25

Two-Way ANOVA Table without Interaction

Source	DF	SS	MS	F	P
Samples	9	282.49	31.388	3.0523	0.003
Operators	4	611.14	152.785	14.8573	0.000
Repeatability	86	884.38	10.283		
Total	99	1778.01			

Gage R&R

Source	VarComp	%Contribution (of VarComp)
Total Gage R&R	17.4086	89.19
Repeatability	10.2835	52.68
Reproducibility	7.1251	36.50
Operators	7.1251	36.50
Part-To-Part	2.1104	10.81
Total Variation	19.5190	100.00

Process tolerance = 15

Source	StdDev (SD)	Study Var (5.15 * SD)	%Study Var (%SV)	%Tolerance (SV/Toler)
Total Gage R&R	4.17236	21.4876	94.44	143.25
Repeatability	3.20679	16.5150	72.58	110.10
Reproducibility	2.66929	13.7468	60.42	91.65
Operators	2.66929	13.7468	60.42	91.65
Part-To-Part	1.45274	7.4816	32.88	49.88
Total Variation	4.41803	22.7529	100.00	151.69

Number of Distinct Categories = 1

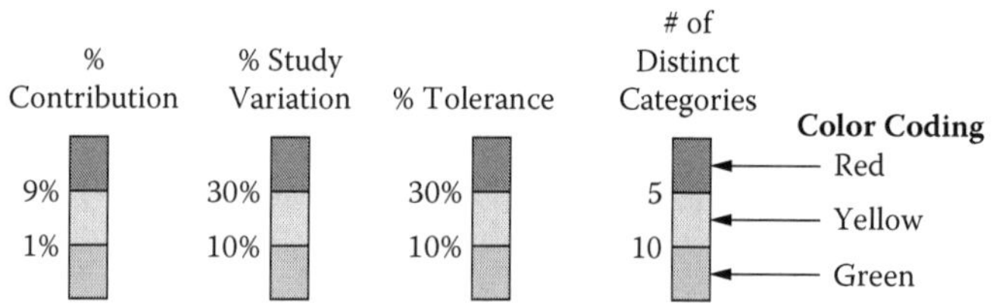

FIGURE 11.44
Gage R&R results.

There is no reproducibility or repeatability between any of the measurements.

Process Capabilities

The capability of a process is the spread that contains most of the values of the process distribution (Figure 11.46). Capability can only be established on a process that is stable with a distribution that only has common cause variation.

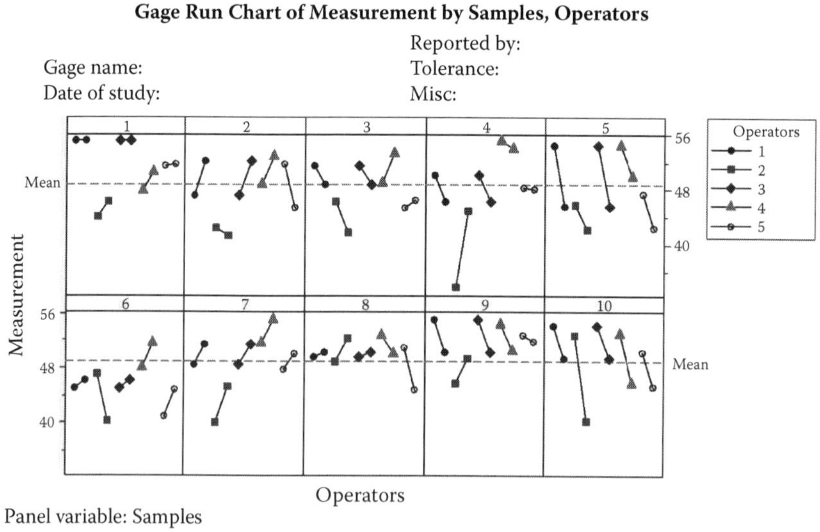

FIGURE 11.45
Gage R&R run chart.

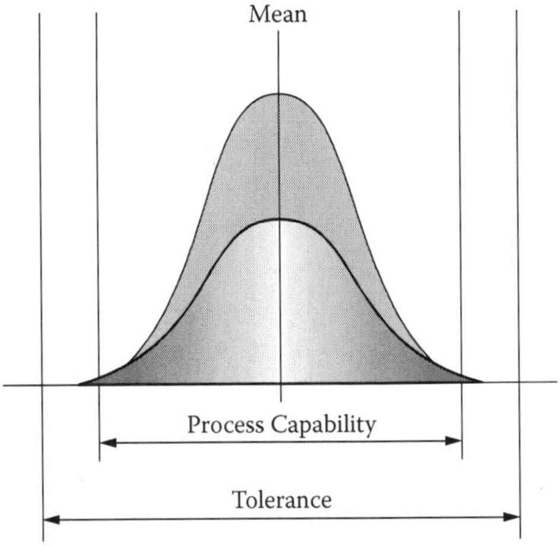

FIGURE 11.46
Process capability.

Process Capability Example: Capable Process (C_p)

A process is capable ($C_p \geq 1$) if its natural tolerance lies within the engineering tolerance or specifications. The measure of process capability of a stable process is $6\hat{\sigma}$, where $\hat{\sigma}$ is the inherent process variability that is estimated from the process. A minimum value of $C_p = 1.33$ is generally used for an ongoing process. This ensures a very low reject rate of 0.007% and therefore is an effective strategy for the prevention of nonconforming items. C_p is defined mathematically as

$$C_p = \frac{USL - LSL}{6\hat{\sigma}}$$

$$= \frac{\text{allowable process spread}}{\text{actual process spread}}$$

where

USL = Upper specification limit
LSL = Lower specification limit

C_p measures the effect of the inherent variability only. The analyst should use R-bar/d_2 to estimate $\hat{\sigma}$ from an R-chart that is in a state of statistical control, where R-bar is the average of the subgroup ranges and d_2 is a normalizing factor that is tabulated for different subgroup sizes (n). We do not have to verify control before performing a capability study. We can perform the study, then verify control after the study with the use of control charts. If the process is in control during the study, then our estimates of capabilities are correct and valid. However, if the process was not in control, we would have gained useful information, as well as proper insights as to the corrective actions to pursue.

Capability Index (C_{pk})

Process centering can be assessed when a two-sided specification is available. If the capability index (C_{pk}) is equal to or greater than 1.33, then the process may be adequately centered. C_{pk} can also be employed when there is only one-sided specification. For a two-sided specification, it can be mathematically defined as

$$C_{pk} = Minimum\left\{\frac{USL - \bar{X}}{3\hat{\sigma}}, \frac{\bar{X} - LSL}{3\hat{\sigma}}\right\}$$

where $\bar{X}$ is the overall process average.

However, for a one-sided specification, the actual C_{pk} obtained is reported. This can be used to determine the percentage of observations out of specification. The overall long-term objective is to make C_p and C_{pk} as large as possible by continuously improving or reducing process variability, $\hat{\sigma}$, for every iteration so that a greater percentage of the product is near the key quality characteristics target value. The ideal is to center the process with zero variability.

If a process is centered but not capable, one or several courses of action may be necessary. One of the actions may be that of integrating a designed experiment to gain additional knowledge on the process and in designing control strategies. If excessive variability is demonstrated, one may conduct a nested design with the objective of estimating the various sources of variability. These sources of variability can then be evaluated to determine what strategies to use to reduce or permanently eliminate them. Another action may be that of changing the specifications or continuing production and then sorting the items. Three characteristics of a process can be observed with respect to capability:

1. The process may be centered and capable.
2. The process may be capable but not centered.
3. The process may be centered but not capable.

Possible Applications of a Process Capability Index

The potential applications of a process capability index are summarized next:

- *Communication*—C_p and C_{pk} have been used in industry to establish a dimensionless common language useful for assessing the performance of production processes. Engineering, quality, manufacturing, and so on, can communicate and understand processes with high capabilities.

- *Continuous Improvement*—The indices can be used to monitor Continuous Improvement by observing the changes in the distribution of process capabilities. For example, if there were 20% of processes with capabilities between 1 and 1.67 in a month, and some of these improved to between 1.33 and 2.0 the next month, then this is an indication that improvement has occurred.
- *Audits*—There are many kinds of audits in use today to assess the performance of quality systems. A comparison of in-process capabilities with capabilities determined from audits can help establish problem areas.
- *Prioritization of improvement*—A complete printout of all processes with unacceptable C_p or C_{pk} values can be extremely powerful in establishing the priority for process improvements.
- *Prevention of nonconforming product*—For process qualification, it is reasonable to establish a benchmark capability of $C_{pk} = 1.33$, which will make nonconforming products unlikely in most cases.

Potential Abuse of C_p and C_{pk}

In spite of its several possible applications, the process capability index has some potential sources of abuse:

- *Problems and drawbacks*—C_{pk} can increase without process improvement even though repeated testing reduces test variability. The wider the specifications, the larger the C_p or C_{pk}, but the action does not improve the process.
- *Analysts tend to focus on number rather than on process.*
- *Process control*—Analysts tend to determine process capability before statistical control has been established. Most people are not aware that capability determination is based on common cause process variation and what can be expected in the future. The presence of special causes of variation makes prediction impossible and capability index unclear.
- *Nonnormality*—Some processes result in nonnormal distribution for some characteristics. Since capability indices are very sensitive to departures from normality, data transformation may be used to achieve approximate normality.
- *Computation*—Most computer-based tools do not use $\bar{R}/d_2$ to calculate σ.

When analytical and statistical tools are coupled with sound managerial approaches, an organization can benefit from a robust implementation of improvement strategies. One approach that has emerged as a sound managerial principle is Lean, which has been successfully applied to many industrial operations.

C_p and C_{pk} are capability analyses that can only be done with normal data. It is very easy to use any data for capability analyses, especially on software systems that will calculate the data automatically. The first step in doing the capability analysis is to have continuous data and check for normality. Only if the data is normal, the capability studies can be done. If the data is not normal, the special cause variation is sought after. Data points may only be taken out if the reasoning is known for the data point that is an outlier (i.e., temperature change, shift change, etc.). Once an outlier is found for a known reason, the outlier can be removed and the data can be checked for normality once again. If there is no root cause for the outlier, more data must be taken, but capability analyses should not be done until the normality is proven.

The importance of finding the capable equipment or products in a business through process capabilities will allow the variation to be found through benchmarking. The best processes should be used for these benchmarking techniques. The Best in Class (BIC) practices should be performed on the different equipment, products, and processes. Then improvements should be made on the areas that are not as capable. It is very important to perform preventative maintenance on all and any equipment in order for the equipment to stay performing at the highest possible process capability.

When two or more equipment pieces are being compared, the first step is to perform a normality test (see Figures 11.47 to 11.51). The conclusions that come from the normality tests are the following:

Blender A1 is not normal.

Blender B1 is the most normal.

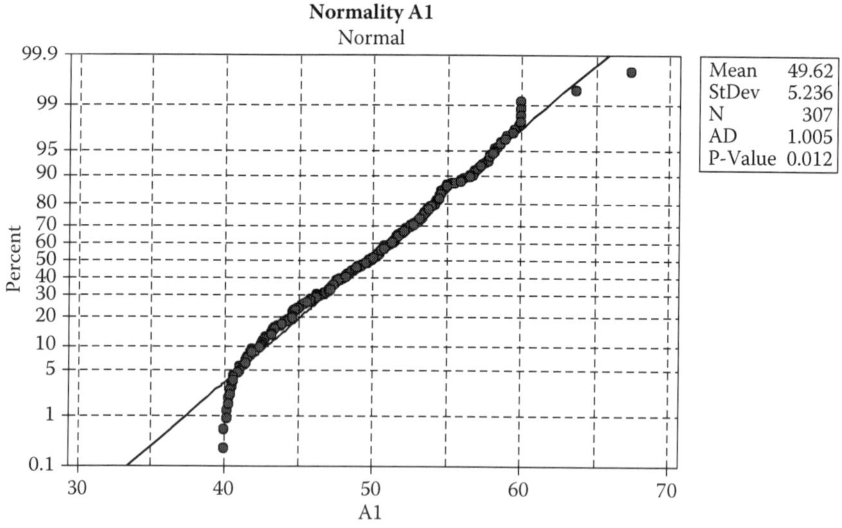

FIGURE 11.47
Normality of Equipment A.

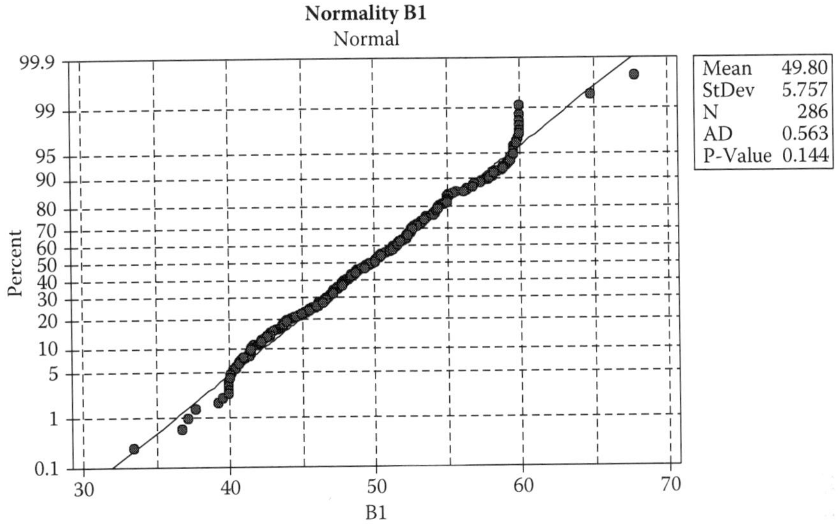

FIGURE 11.48
Normality of Equipment B1.

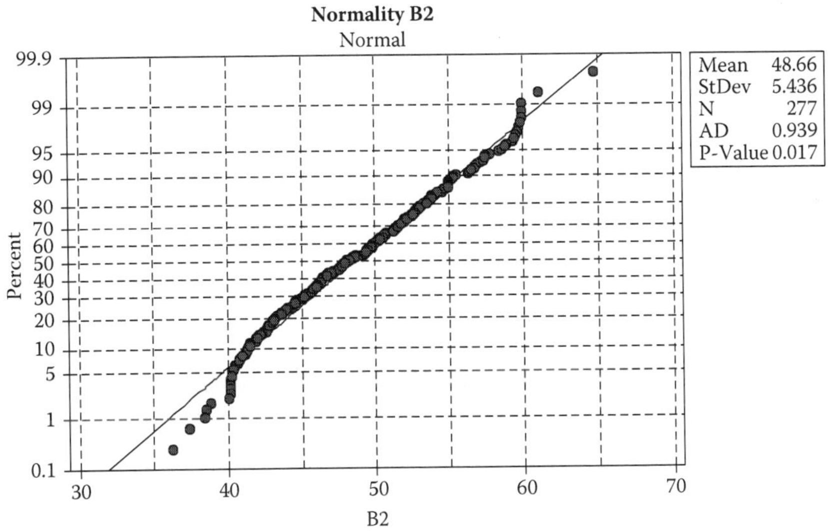

FIGURE 11.49
Normality of Equipment B2.

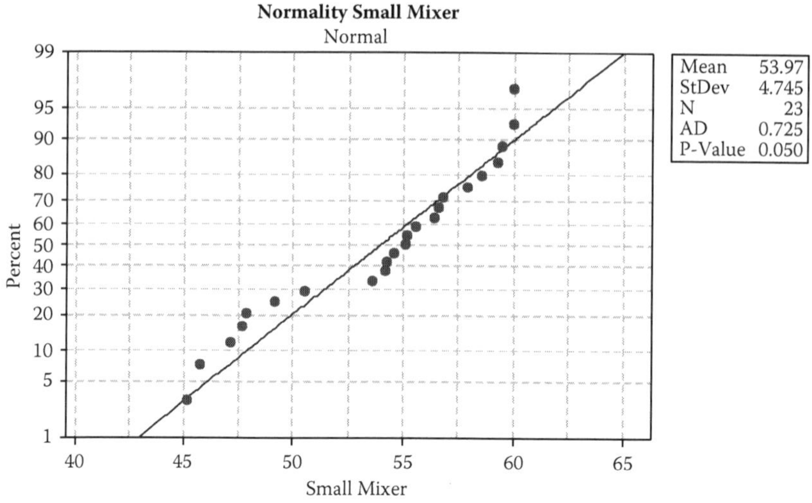

FIGURE 11.50
Normality of Equipment Small Mixer.

Blender B2 is not normal.

Small Mixer is just in at normal.

The normal pieces of equipment are then checked for the process capabilities (see Figure 11.51 and Figure 11.52).

Blender B1 is your best blender, the short-term capability (C_p of 1.34) is approximately equivalent to a short-term Z of 4, which is good. The long-term capability needs some improvement (P_{pk} of 0.94).

Small Mixer is just normal, but is better than Blender A1 or B2. The short-term capability (C_p of 0.69) needs improvement along with the long-term capability (P_{pk} of 0.42).

The process capability analyses should be continued in a systematic fashion (i.e., monthly or quarterly) to understand if the processes are improving. Continuous Improvement should be performed on the equipment for the best capabilities.

Variation

Variation is present in all processes, but the goal is to reduce the variation while understanding the root cause of where the variation comes from in the process and why. For Six Sigma to be successful, the processes must be in control statistically and must be improved by reducing the variation. The distribution of the measurements should be analyzed to find the variation and depict the outliers or patterns.

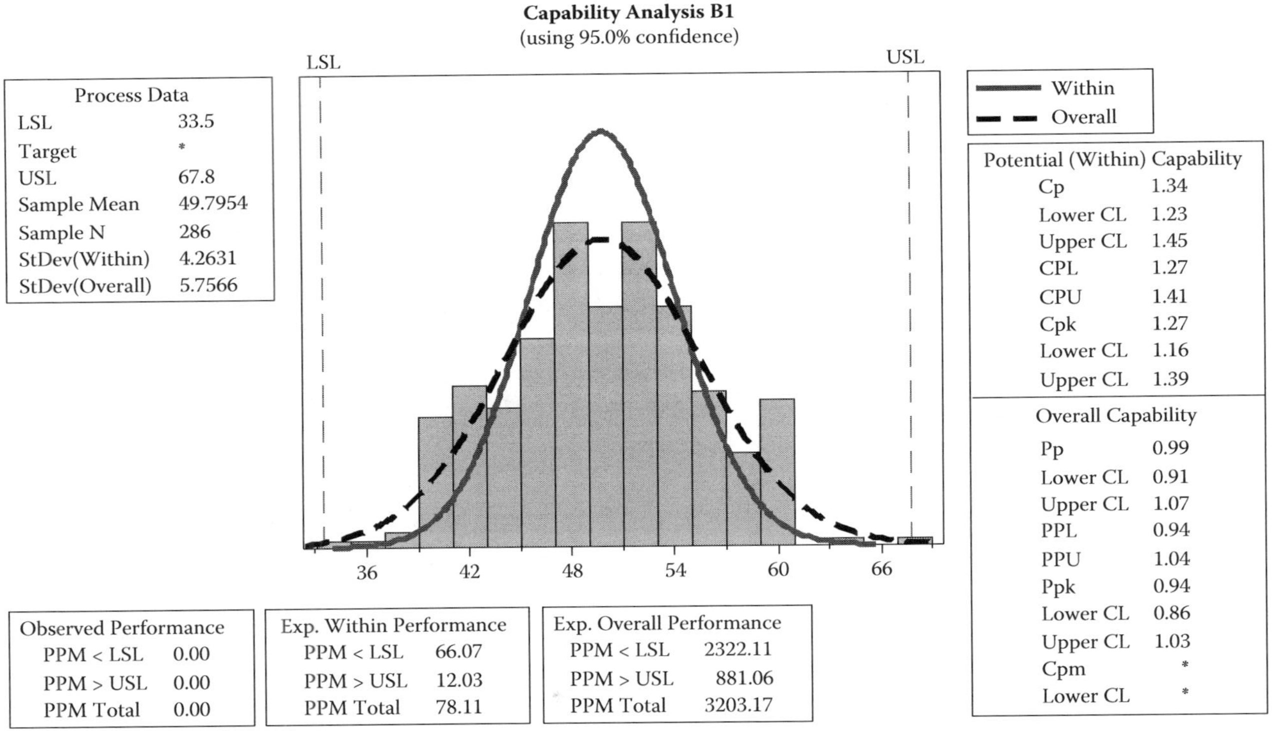

FIGURE 11.51
Process capability B1.

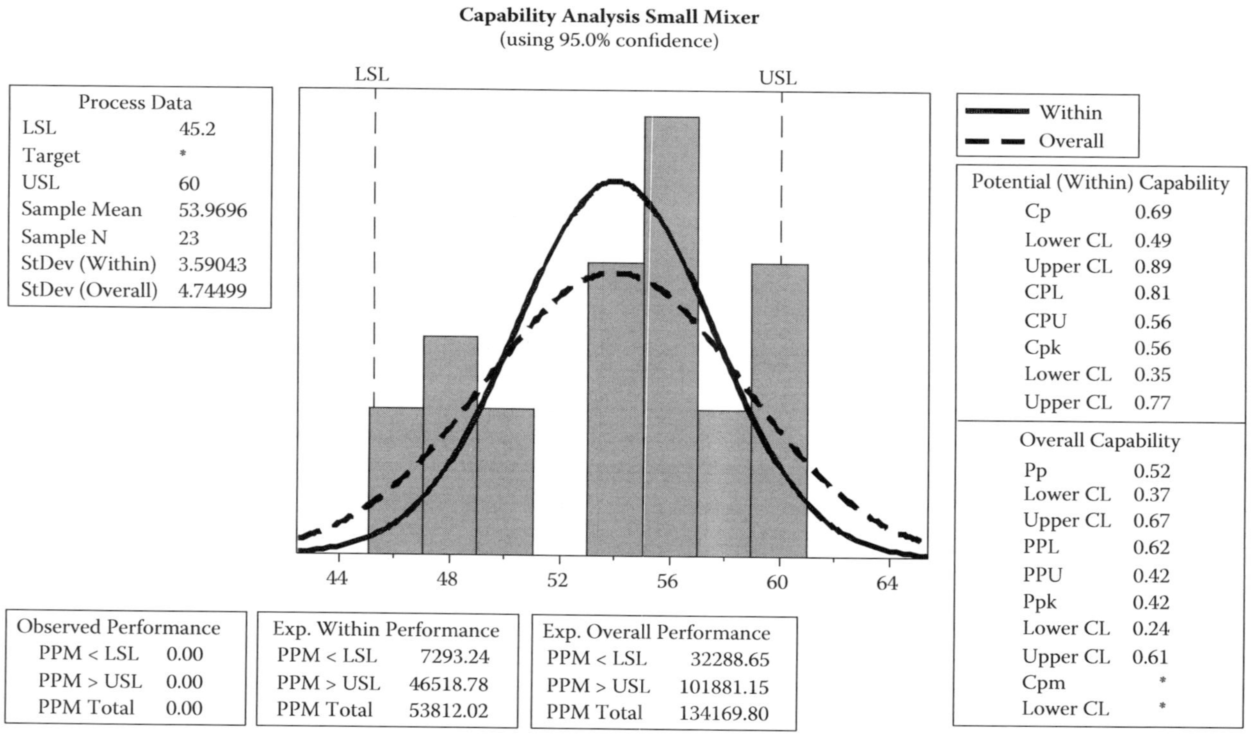

FIGURE 11.52
Process capability Small Mixer.

The study of variation began with Dr. W. E. Deming, who was also known as the Father of Statistics. Deming stated that variation happens naturally, but the purpose is to utilize statistics to show patterns and types of variations. There are two types of variations that are sought after: special cause variation and common cause variation. Special cause variation refers to out-of-the-ordinary events, such as a power outage, whereas common cause variation is inherent in all processes and is typical. The variation is sought to be reduced so that the processes are predictable, in statistical control, and have a known process capability. A root cause analysis should be done on special cause variation so that the occurrence is not to happen again. Management is in charge of common cause variation where action plans are given to reduce the variation.

Assessing the location and spread are also important factors. Location is known as the process being centered along with the process requirements. Spread is known as the observed values compared to the specifications. The stability of the process is required. The process is said to be in statistical control if the distribution of the measurements have the same shape, location, and spread over time. This is the point in time where all special causes of variation are removed and only common cause variation is present.

An *average, central tendency* of a data set is a measure of the "middle" or "expected" value of the data set. Many different descriptive statistics can be chosen as measurements of the central tendency of the data items. These include the arithmetic mean, the median, and the mode. Other statistical measures such as the standard deviation and the range are called measures of spread of data. An average is a single value meant to represent a list of values. The most common measure is the arithmetic mean but there are many other measures of central tendency such as the median (used most often when the distribution of the values is skewed by small numbers with very high values).

As stated before, special cause variation would be occurrences such as power outages and large mechanical breakdowns. Common cause variations would be occurrences such as electricity being different by a few thousand kilowatts per month. To understand the variation, graphical analyses should be done followed by capability analyses.

It is important to understand the variation in the systems so that the best performing equipment is used. The variation sought after is in turn utilized for sustainability studies. The best performing equipment should be utilized the most and the least performing equipment should be brought back to its original state of condition and then upgraded or fixed to be capable. Capability indices are explained next.

Graphical Analysis

Graphical analyses are visual representations of tools that show meaningful key aspects of projects. These tools are commonly known as dotplots, histograms, normality plots, Pareto diagrams, second level Paretos (also known

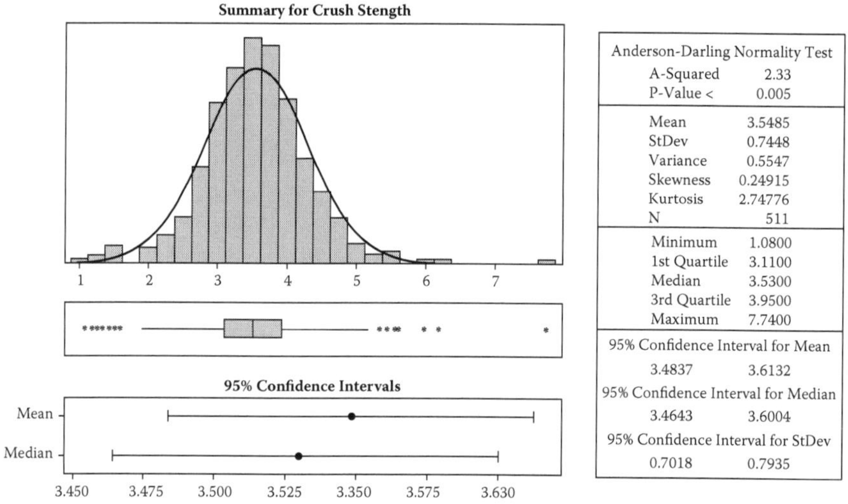

FIGURE 11.53
Graphical analysis.

as stratification), boxplots, scatter plots, and marginal plots. The plotting of data is a key beginning step to any type of data analysis because it is a visual representation of the data. A graphical analysis is shown in Figure 11.53.

A graphical analysis summary can cover sample size, mean, standard deviation, variance, skewness, kurtosis, p-value, and confidence intervals, as shown in Figure 11.54.

A Pareto chart will be able to show a visual representation of what occurs the most, as shown in Figure 11.55.

Cause and Effect Diagram

After a process is mapped, the cause and effect (C&E) diagram can be completed. This process is important because it is a completed root cause analysis. The basis behind root cause analysis is to ask why five times to get to the actual root cause. Many times, problems are Band-Aided to fix the top-level problem, but the actual problem itself is not addressed.

A cause and effect diagram is shown in Figure 11.55. The cause and effect diagram is also referred to as a fishbone diagram because visually it looks like a fish where the bones are the causes and the fish head is the effect. The fishbone is broken out to the most important categories in an environment:

- Measurements
- Material

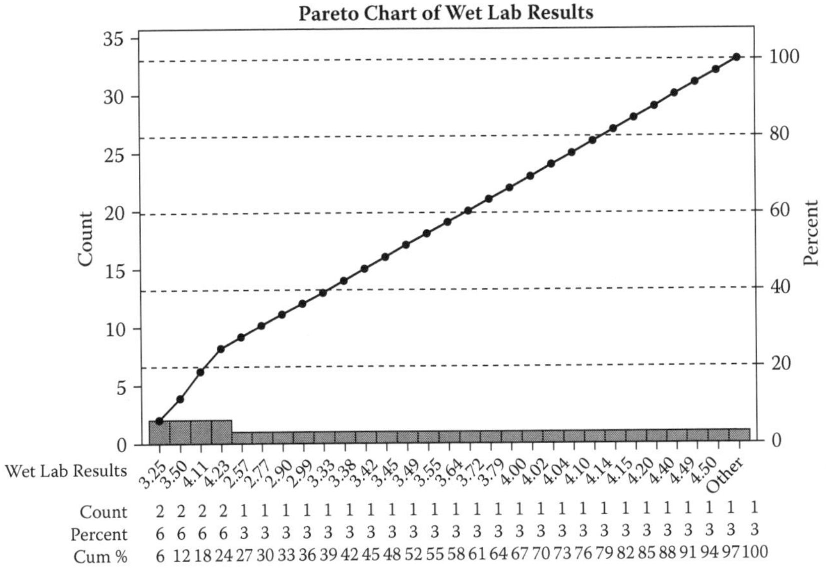

FIGURE 11.54
Pareto chart.

- Personnel
- Environment
- Methods
- Machines

This process requires a team to do a great deal of brainstorming where it focuses on the causes of the problems based on the categories. The "fish head" is the problem statement.

Failure Mode and Effect Analysis (FMEA)

In order to select action items from the cause and effect diagram and prioritize the projects, failure mode and effect analyses are completed. The FMEA will identify the causes, assess risks, and determine further steps. The steps to an FMEA are the following:

1. Define process steps
2. Define functions
3. Define potential failure modes
4. Define potential effects of failure

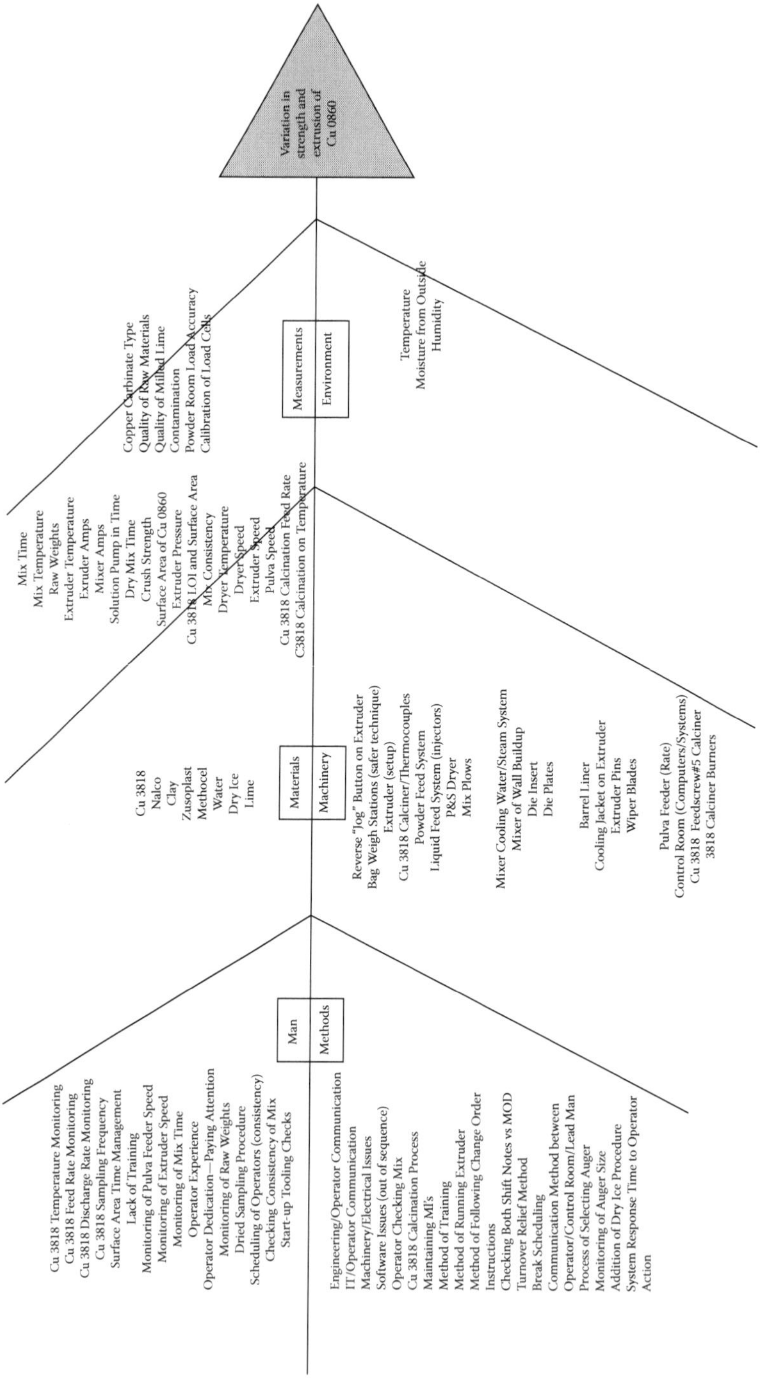

FIGURE 11.55
Cause and effect diagram.

5. Define the severity of a failure
6. Define the potential mechanisms of failure
7. Define current process controls
8. Define the occurrence of failure
9. Define current process control detection mechanisms
10. Define the ease of detecting a failure
11. Multiply severity, occurrence, and detection to calculate a risk priority number (RPN)
12. Define recommended actions
13. Assign actions with key target dates to responsible personnel
14. Revisit the process after actions have been taken to improve it
15. Recalculate RPNs with the improvements

The detection, severity, and occurrence criteria are shown in Tables 11.4 to 11.6.

An FMEA is shown in Figure 11.56. The FMEA shows that an important aspect to sustainability is the reduction of the RPN after the action items. It is important to understand the process severity to a customer and increasing the capability of the process to in turn improve the process. The reducing RPN will make the entire process more sustainable by being able to deliver the process at the best capabilities through thorough project management. It is important to maintain the FMEA so that once a process is improved it is not forgotten.

TABLE 11.4

Detection Criteria

Detection	Criteria: Likelihood That the Existence of a Defect Will Be Detected by Test Content before the Product Advances to the Next or Subsequent Process	Ranking
Almost impossible	Test content detects < 80% of failures	10
Very remote	Test content must detect < 80% of failures	9
Remote	Test content must detect < 82.5% of failures	8
Very low	Test content must detect < 85% of failures	7
Low	Test content must detect < 87.5% of failures	6
Moderate	Test content must detect < 90% of failures	5
Moderately high	Test content must detect < 92.5% of failures	4
High	Test content must detect < 95% of failures	3
Very high	Test content must detect < 97.5% of failures	2
Almost certain	Test content must detect < 99.5% of failures	1

TABLE 11.5

Severity Criteria

Effect	Criteria: Severity of the Effect	Ranking
Hazardous, without warning	Very high severity ranking when a potential failure mode affects safety and involves noncompliance without warning	10
Hazardous, with warning	Very high severity ranking when a potential failure mode affects safety and involves noncompliance with warning	9
Very high	Process is not operable and has loss of its primary function	8
High	Process is operable, but with a reduced functionality and an unhappy customer	7
Moderate	Process is operable, but not easy to manufacture; customer is uncomfortable	6
Low	Process is operable but uncomfortable with a reduced level of performance; customer is dissatisfied	5
Very low	The process is not in 100% compliance; most customers are able to notice the defect	4
Minor	The process is not in 100% compliance; some customers are able to notice the defect	3
Very minor	The process is not in 100% compliance; very few customers are able to notice the defect	2
None	No effect	1

TABLE 11.6

Occurrence Criteria

Probability of Failure	Possible Failure Rates	Ranking
Failures are almost inevitable	≥1 in 2	10
	1 in 3	9
High: Repeated failures	1 in 8	8
	1 in 20	7
Moderate: Occasional failures	1 in 80	6
	1 in 400	5
	1 in 2000	4
Low: Very few failures	1 in 15,000	3
	1 in 150,000	2
Remote: Failure is unlikely	≤1 in 1,500,000	1

Process Function (Step)	Potential Failure Modes (process defects)	Potential Failure Effects (KPOVs)	SEV	Class	Potential Causes of Failure (KPIVs)	OCC	Current Process Controls	DET	RPN	Recommend Actions	Responsible Person & Target Date	Taken Actions	SEV	OCC	DET	RPN
Copper Strikes	Agitator	100% Down, potential over reactor	9		Motor, Bearing, Shaft, Gearbox	3	None	5	135							
APV	Product Collector Roto Lock	Product not discharging	8		Motor bad, Communication with scale, jams	4	None	2	64							
APV	Blowers	100% Down	8		Plugged filter, bad motor, bad coupling	3	PM on blowers every 4 months	2	48							
Copper nitrate makeup	Discharge valve	Pluggage	8		Powder build up before dissolved	2	Valve design – flush mount	3	48							
Copper nitrate makeup	Mag drive pump	Won't pump	8		Running dry, wom out, motor failure	6	Level indication, recirculation	1	48	Load monitor to be put on redundant pump						
Copper nitrate makeup	Gate failure	100% Down	8		Damage	6	None	1	48	Limit switch, investigate new gate						

FIGURE 11.56
Failure modes and effects matrix.

Hypothesis Testing

Hypothesis testing validates assumptions made by verification of the processes based on statistical measures. It is important to use at least 30 data points for hypothesis testing so that there is enough data to validate the results. Normality of the data points must be found in order for the hypothesis testing to be accurate. The assumptions are shown in the null and alternate hypotheses:

H_0 (null hypothesis)—The difference is equal to the chosen reference value $\mu_1 - \mu_2 = 0$.

H_a (alternate hypothesis)—The difference is not equal to the chosen reference value $\mu_1 - \mu_2$ is not = 0.

An example of a paired t-test for hypothesis testing is shown in Table 11.7.

95% CI for mean difference: (1.16, 6.69)

t-test of mean difference = 0 (versus not = 0)

t-value = 2.90

p-value = 0.007

The confidence interval for the mean difference between the two materials does not include zero, which suggests a difference between them. The small p-value ($p = 0.007$) further suggests that the data are inconsistent with H_0 (null hypothesis). The difference is not equal to the chosen reference value and therefore, $\mu_1 - \mu_2$ is not = 0. Specifically, the first set (mean = 79.697) performed better than the next set (mean = 83.623) in terms of weight control over the time span. Conclusion, reject H_0, the difference is not equal to the chosen reference value: $\mu_1 - \mu_2$ is not = 0. The histogram of differences and the boxplot of differences are graphed in Figures 11.57 and 11.58, respectively.

ANOVA

The purpose of an ANOVA, also known as analysis of variance, is to determine if there is a relationship between a discrete, independent variable, and

TABLE 11.7

Hypothesis Testing Paired t-Test Example

Paired T for Before–After	N	Mean	StDev	SE Mean
Before	30	83.623	5.195	0.948
After	30	79.697	4.998	0.913
Difference	30	3.93	7.41	1.35

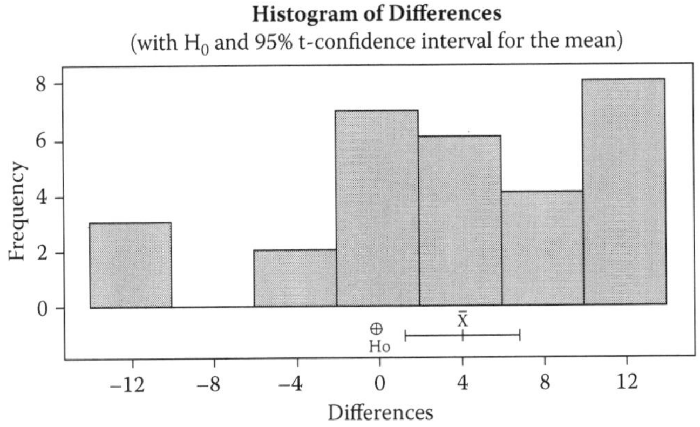

FIGURE 11.57
Histogram of differences.

a continuous, dependent output. There is a one-way ANOVA, which includes one-factorial variance, and a two-way ANOVA, which includes a two-factorial variance. Three sources of variability are sought after:

Total—Total variability within all observations

Between—Variation between subgroup means

Within—Random chance variation within each subgroup, also known as noise

The equation for a one-way ANOVA is

$$SS_T = SS_F + SS_e$$

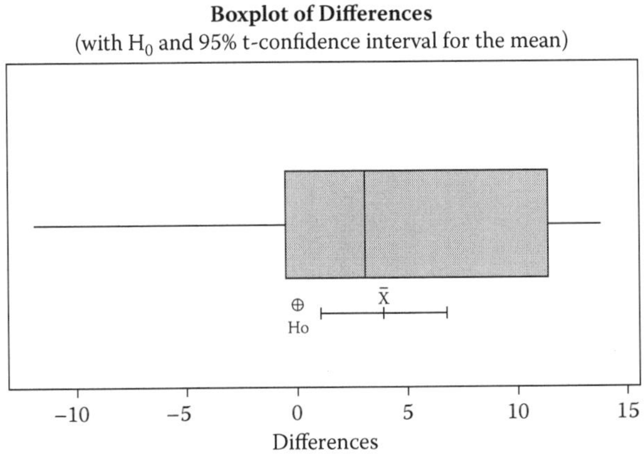

FIGURE 11.58
Boxplot of differences.

The principles for the one-way ANOVA and two-way ANOVA are the same except that in a two-way ANOVA, the factors can take on many levels. The total variability equation for a two-way ANOVA is

$$SS_T = SS_A + SS_B + SS_{AB} + SS_e$$

where

SS_T = Total sum of squares
SS_F = Sum of squares of the factor
SS_e = Sum of squares from error
SS_A = Sum of squares for factor A
SS_B = Sum of squares for factor B
SS_{AB} = Sum of squares due to interaction of factors A and B

If the ANOVA shows that at least one of the means is different, a pairwise comparison is done to show which means are different. The residuals, variance, and normality should be examined, and the main effects plot and interaction plots should be generated.

The F-ratio in an ANOVA compares the denominator to the numerator to see the amount of variation that is expected. When the F-ratio is small, which is normally close to 1, the value of the numerator is close to the value of the denominator, and the null hypothesis cannot be rejected stating the numerator and denominator are the same. A large F-ratio indicates the numerator and denominator are different also known as the MS error where the null hypothesis is rejected.

Outliers should also be sought after in the ANOVA showing the variability is affected.

The main effects plot shows the mean values for the individual factors being compared. The differences between the factor levels can be seen with the slopes in the lines. The p-values can help determine if the differences are significant.

Interaction plots show the mean for different combinations of factors.

Correlation

The linear relationship between two continuous variables can be measured through correlation coefficients. The correlation coefficients are values between –1 and 1.

If the value is around 0, there is no linear relationship.

If the value is less than .05, there is a weak correlation.

If the value is less than .08, there is a moderate correlation.

If the value is greater than .08, there is a strong correlation.

If the value is around 1, there is a perfect correlation.

Simple Linear Regression

The regression analysis describes the relationship between a dependent and independent variable as a function y = f(x). The equation for simple linear regression as a model is

$$Y = b_0 + b_1 x + {}_E$$

where

Y = dependent variable

b_0 = axis intercept

b_1 = gradient of the regression line

x = independent variable

$_E$ = error term or residuals

The predicted regression function is tested with the following formula:

$$R^2 = \frac{SSTO - SSE}{SSTO}$$

where

$$SSTO = \begin{cases} \underset{\sim}{Y}' - n\bar{Y}^2 & \text{if constant} \\ \underset{\sim}{Y}\underset{\sim}{Y} & \text{if no constant} \end{cases}$$

Note: When the no constant option is selected, the total sum of square is uncorrected for the mean. Thus, the R^2 value is of little use, since the sum of the residuals is not zero.

The F-test shows if the predicted model is valid for the population and not just the sample. The model is statistically significant if the predicted model is valid for the population.

The regression coefficients are tested for significance through t-tests with the following hypothesis:

H_0: $b_0 = 0$, the line intersects the origin

H_a: $b_0 \neq 0$, the line does not intersect the origin

H_0: $b_1 = 0$, there is no relationship between the independent variable xi, and the dependent variable y

H_a: $b_1 \neq 0$, there is a relationship between the independent variable xi, and the dependent variable y

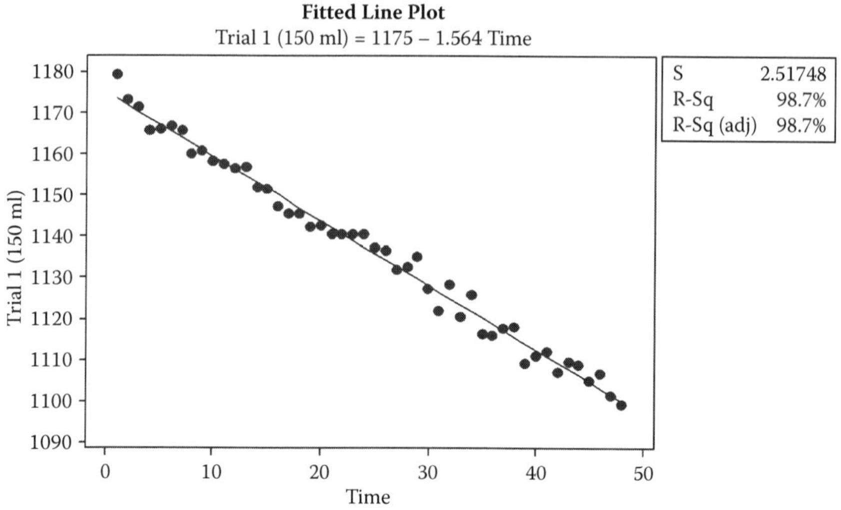

FIGURE 11.59
Fitted line plot.

A fitted line plot can be done to see the inverse relationship, as shown in Figure 11.59. After the inverse relationship is seen, a regression analysis can be performed. An example for the analysis of whether there was a pressure degradation over time on a particular piece of equipment is explained next. A linear relationship was sought after. First, it was sought to see if there was correlation since it can be seen that there is a linear relationship between the variables. The y was the measurement and the x was the time.

Correlations: Trial 1, Time

Pearson correlation of Trial 1 (150 ml) and Time = –0.994

p-value = 0.000

This correlation coefficient of r = –0.994 shows a high, positive dependence.

The p-value being less than .05 also shows that the correlation coefficient is significant.

The regression equation is

$$\text{Trial 1} = 1175 - 1.56\ \text{Time}$$

Predictor	**Coef**	**SE Coef**	**T**	**P**
Constant	1174.89	0.74	1591.47	0.000
Time	–1.56360	0.02623	–59.61	0.000

S = 2.51748; R-Sq = 98.7%; R-Sq(adj) = 98.7%

Analysis of Variance

Source	DF	SS	MS	F P
Regression 1	22522	22522	3553.63	0.000
Residual Error	46	292	6	
Total	47 22813			

Unusual Observations

Trial 1

Obs	Time	(150 ml)	Fit	SE Fit	Residual	St Resid
1	1.0	1179.00	1173.33	0.72	5.67	2.35R
29	29.0	1135.10	1129.55	0.38	5.55	2.23R

Note: R denotes an observation with a large standardized residual.

For each time, there is 1.5 measurement of degradation according to the equation

$$Y_1 = \beta o + \beta_1 X_1$$

The slope equals 1.564. There is a negative slope.

In this particular situation, it becomes critical after losing more than 10% of the measurement in the specification range. According to the graph, in about every 10 trials, there is a degradation of about 15.

Hypothesis Testing

The p-value shows that this is not normal.

H_0 = Accept null hypothesis ($\beta = 0$, no correlation)

H_a = Reject null hypothesis ($\beta \neq 0$, there is correlation)

Therefore, reject null hypothesis. There is correlation.

In the above, the R^2 value is 98.7%, which means 98.7% of the Y variable's pressure can be explained by the model (the regression equation).

The residuals are then evaluated, as shown in Figure 11.60. The normality is also taken of the residuals (Figure 11.61). The normality test passes with a value of .850. The residuals are in control. The residuals are contained in a straight band, with no obvious pattern in Figure 11.62, showing that this model is adequate.

Conclusion: Reject H_0, the slope of the line does not equal 0. There is a linear relationship in the measurement versus time, showing that there is correlation. This model proves to be adequate.

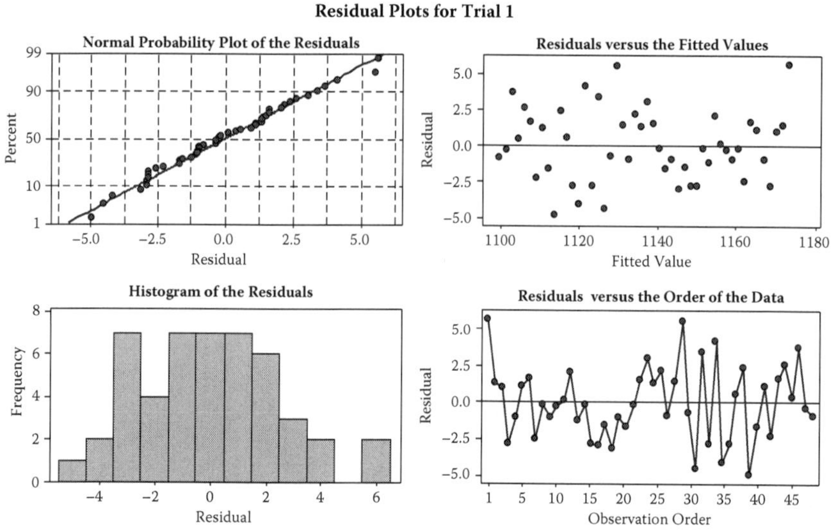

FIGURE 11.60
Residual plots.

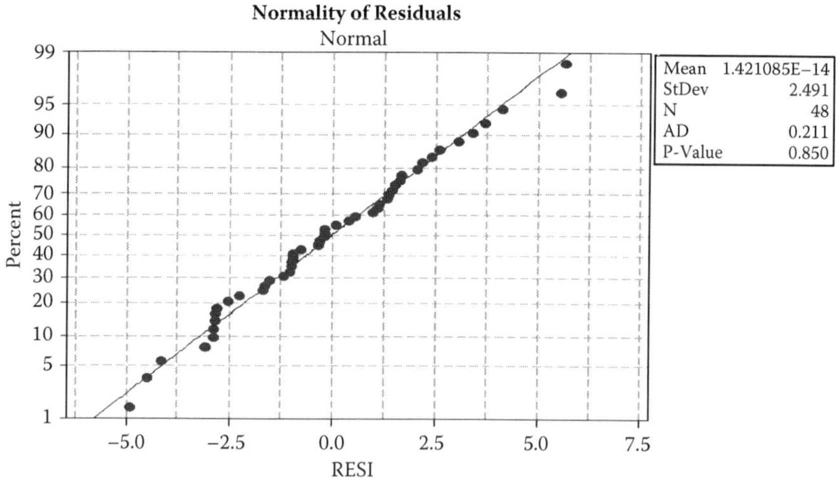

FIGURE 11.61
Normality of residuals.

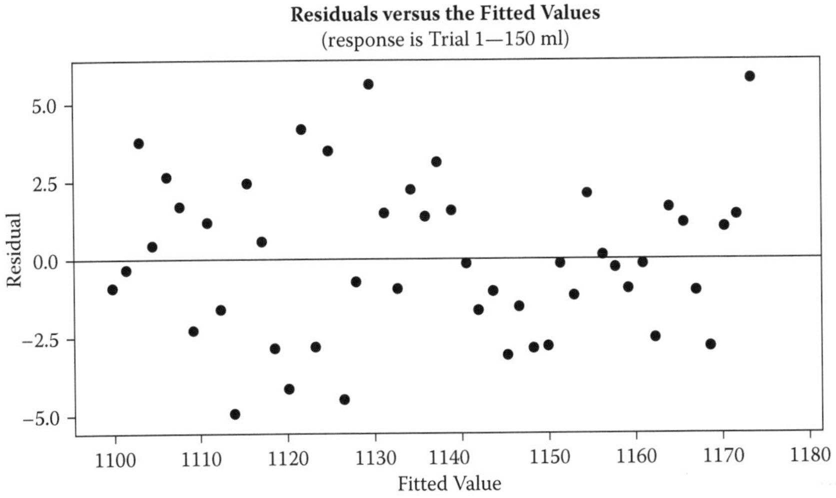

FIGURE 11.62
Residuals versus fitted values.

Theory of Constraints

Dr. Eliyahu M. Goldratt created a theory of constraints (TOC). This management theory proved that every system has at least one constraint limiting it from 100% efficiency. The analysis of a system will show the boundaries of the system. TOC not only shows the cause of the constraints, but it also provides a way to resolve the constraints. There are two underlying concepts with TOC:

1. System as Chains
2. Throughput, Inventory Management, and Operating Expenses

The performance of the entire system is called the chain. The performance of the system is based on the weakest link of the chain or the constraint. The remaining links are known as nonconstraints. Once the constraint is improved, the system becomes more productive or efficient, but there is always a weakest link or constraint. This process continues until there is 100% efficiency.

If there are three manufacturing lines and they produce the following:

1. 250 units/day
2. 500 units/day
3. 600 units/day

The weakest link is manufacturing line 1 because it produces the least amount of units per day. The weakest link is investigated until it reaches the capacity of the nonconstraints. After the improvement has been made, the new weakest link is investigated until the full potential of the manufacturing lines can be fulfilled without exceeding market demand. If the external demand is fewer than the internal capacity, it is known as an external constraint.

Throughput can be defined in the following formula:

(Sales Price – Variable Costs)/Time

Profits should be understood when dealing with throughput.

Inventories are known as raw materials, unfinished goods, purchased parts, and investments made. Inventory should be seen as dollars on shelves. Any inventory is a waste unless utilized in a just-in-time manner.

Operating expenses should include all expenses utilized to produce a good. The less the operating expenses, the better. These costs should include direct labor, utilities, supplies, and depreciation of assets.

Applying the TOC concept helps guide making the weakest link stronger. There are five steps to the process of TOC:

1. Identify the constraint or the weakest link.
2. Exploit the constraint by making it as efficient as possible without spending money on the constraint or considering upgrades.
3. Subordinate everything else to the constraint; adjust the rest of the system so the constraint operates at its maximum productivity. Evaluate the improvements to ensure the constraint has been properly addressed and it is no longer the constraint. If it is no longer the constraint, skip step 5.
4. Elevate the constraint. This step is only required if steps 2 and 3 were not successful. The organization should take any action on the constraint to eliminate the problem. This is the process where money should be spent on the constraint or upgrades should be investigated.
5. Identify the next constraint and begin the 5-step process over. The constraint should be monitored and Continuous Improvement should be completed.

Single Minute Exchange of Dies (SMED)

Single Minute Exchange of Dies (SMED) is a theory and set of techniques that make it possible to perform equipment setup and changeover operations in under 10 minutes. It was originally developed to improve die press and machine tool setups but the principles apply to changeovers in all processes. It may not be possible to reach the "single-minute" range for all setups, but SMED dramatically reduces setup times in almost every case. SMED leads to benefits for the company by giving customers a variety of products in just the quantities they need. SMED can result in high quality, good price, speedy delivery, less waste, and cost efficiency.

It is important to understand large lot production, which leads to trouble. The three key topics to consider when understanding large lot production are:

- Inventory Waste
  - Storing what is not sold costs money
  - Ties up company resources with no value to the product

- Delay
  - Customers have to wait for the company to produce entire lots rather than just what they want

- Declining Quality
  - Storing unsold inventory increases chances of product being scrapped or reworked, adding costs

Once this is realized, the benefits of SMED can be understood:

- Flexibility
  - Meet changing customer needs without excess inventory

- Quicker delivery
  - Small-lot production equals less lead time and less customer waiting time

- Better quality
  - Less inventory storage equals fewer storage-related defects
  - Reduction of setup errors and elimination of trial runs for new products

- Higher productivity
  - Reduction in downtime
  - Higher equipment productivity rate

Two types of operations are realized during setup operations, which consist of internal and external operations. Internal setup is a setup that can

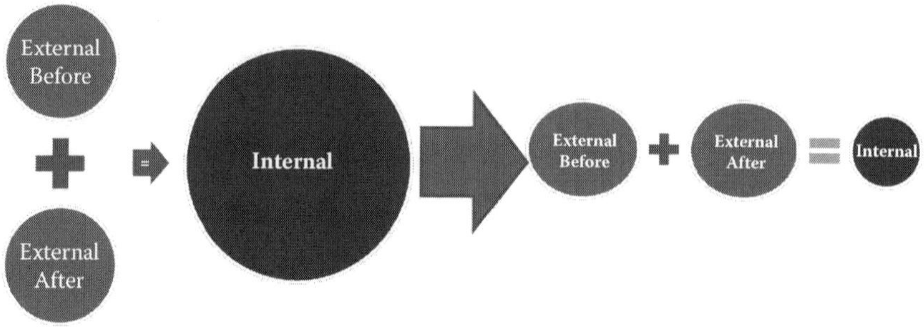

FIGURE 11.63
Internal versus external setup.

only be done when the machine is shut down (i.e., a new die can only be attached to a press when the press is stopped). External setup is a setup that can be done while the machine is still running (i.e., bolts attached to a die can be assembled and sorted while the press is operating).

It is important to convert as much internal work as possible to external work, which is shown in Figure 11.63.

There are four important questions to ask yourself when understanding SMED:

1. How might SMED benefit your factory?
2. Can you see SMED benefiting you?
3. What operations are internal operations?
4. What operations are external operations?

There are three stages to SMED:

1. Separate internal and external setups
   - Distinguish internal versus external
   - By preparing and transporting while the machine is running can cut changeover times by as much as 50%

2. Convert internal setups to external setups
   - Reexamine operations to see whether any steps are wrongly assumed as internal steps
   - Find ways to convert these steps to external setups

3. Streamline all aspects of setup operations
   - Analyze steps in detail
   - Use specific principles to shorten time needed especially for steps internally with machine stopped

Five traditional setups steps are also defined:

Preparation—Ensures that all the tools are working properly and are in the right location

Mounting and extraction—Involves the removal of the tooling after the production lot is completed and the placement of the new tooling before the next production lot

Establishing control settings—Setting all the process control settings prior to the production run; including calibrations and measurements needed to make the machine and tooling effectively operate

First-run capability—This includes the necessary adjustments (recalibrations, additional measurements) required after the first trial pieces are produced

Setup improvement—The time after processing during which the tooling and machinery is cleaned, identified, and tested for functionality prior to storage

In order to determine the proportion of current setup times, the chart in Table 11.8 can be completed.

The three stages of SMED are explained next.

Description of Stage 1: Separate Internal versus External Setup

Three techniques help us separate internal versus external setup tasks:

1. Use checklists
2. Perform function checks
3. Improve transport of die and other parts

TABLE 11.8

Proportion of Setup Times before SMED Improvements

Setup Steps	Setup Type Traditional Internal	Setup Type Traditional External	Resource Consumption (%)	Setup Type One Step Internal	Setup Type One Step External
Preparation	X		20%		X
Mounting and extraction	X		5%	X	
Establish control settings	X		15%		X
First run capability	X		50%	N/A	N/A
Process improvement	X		10%		X

A checklist lists everything required to set up and run the next operation. The list includes items such as:

- Tools, specifications, and workers required
- Proper values for operating conditions such as temperature, pressure, and so forth
- Correct measurement and dimensions required for each operation
- Checking items on the list before the machine is stopped helps prevent mistakes that can occur after internal setup has begun

An operation checklist is shown in Table 11.9.

Function checks:

- Should be performed before setup begins so that repairs can be made if something does not work right.
- If broken dies, molds, or jigs are not discovered until test runs are done, a delay will occur in internal setup.
- Making sure such items are in working order before they are mounted will cut setup time a great deal.

Improve transport of parts and tools:

- Dies, tools, jigs, gauges, and other items needed for an operation must be moved between storage areas and machines, then back to storage once a lot is finished.
- To shorten the time that the machine is shut down, transport of these items should be done during external setup.
- In other words, new parts and tools should be transported to the machine before the machine is shutdown for changeover.

Description of Stage 2: Convert Internal Setups to External Setups

1. Advance preparation of conditions
    - Get the necessary parts, tools, and conditions ready before the internal setup begins.
    - Conditions like temperature, pressure, or position of material can be prepared externally while the machine is running (i.e., preheating of mold/material).

2. Function standardization
    - It would be expensive and wasteful to make external dimensions of every die, tool or part the same, regardless of the size or shape of the product it forms. Function standardizations avoid this

TABLE 11.9

Operation Checklist

Operation Checklist					**Effective:**	**Nov-11**	
Equipment:							
Operation:							
Date:							
		Employees Trained for Setup and Operations (Need 2 people)					
		Name of Employee	x	Name of Employee			
x		Name of Employee		Name of Employee			
		Tools Needed					
		Automatic Nut Driver					
x		Hex Wrench					
x		Rolling Cart					
		Tool					
		Tool					
		Tool					
		Tool					
x		Tool					
		Parts Needed					
x		Elevator Plate – 3.5 lb. Size					
x		Compression Plate – 3.5 lb. Size					
x		Feed Auger					
		Part					
		Part					
		Part					
		Part					
		Standard Operating Procedure to follow					
x		SOP 001 - Changeover Procedure					
x		SOP 003 - Cleandown					
		Procedure					
		Procedure					
		Procedure					
		Procedure					
		Procedure					
		Procedure					

waste by focusing on standardizing only those elements whose functions are essential to the setup.
- Function standardization might apply to dimensioning, centering, securing, expelling, or gripping.

3. Implementing function standardization with two steps:
   - Look closely at each individual function in your setup process and decide which functions, if any, can be standardized.
   - Look again at the functions and think about which can be made more efficient by replacing the fewest possible parts (i.e., clamping function standardization).

Internal versus external setups can be placed in Table 11.10.

TABLE 11.10

Internal versus External Setup Table

INTERNAL VS EXTERNAL SETUPS			
Classify Items under Each Category			
	Internal		External
1		1	
2		2	
3		3	
4		4	
5		5	
6		6	
7		7	
8		8	
9		9	
10		10	
Which items would you convert from internal to external setup?			
1		1	
2		2	
3		3	
4		4	
5		5	
Why			
1			
2			
3			
4			
5			

Description of Stage 3: Streamline All Aspects of the Setup Operation

External setup improvement includes streamlining the storage and transport of parts and tools. When dealing with small tools, dies, jigs, and gauges, it is vital to address the issue of tool and die management. Ask questions like:

- What is the best way to organize these items?
- How can we keep these items maintained in perfect condition and ready for the next operation?
- How many of these items should we keep in stock?

Operation for storing and transporting dies can be very time-consuming, especially when your factory keeps a large number of dies on hand. Storage and transport can be improved by marking the dies with color codes and location numbers on the shelves where they are stored.

To streamline internal setup, implement parallel operations, using functional clamps, eliminating adjustments, and mechanization. Machines such as plastic molding machines and die casting machines often require operation at both the front and back of the machine. One-person changeovers of such machines mean wasted time and movement because the same person is constantly walking back and forth from one end of the machine to the other. Parallel operations divide the setup operation between two people, one at each end of the machine. When setup is done using parallel operations, it is important to maintain reliable and safe operations and minimize waiting time. To help streamline parallel operations, workers should develop and follow procedural charts for each setup.

A setup conversion matrix is shown in Table 11.11.

The final understanding of SMED comes from basic principles such as observing recorded videos of workstations. If there is nobody on the screen, it means there is waste present.

It is important to understand that SMED is more than just a series of techniques. It is a fundamental approach to improving activities. A personal action plan should be found to adhere to each business's needs. It is important to find ways to implement SMED into environments to continue the sustainability of the business. To begin the process, a communication plan should be implemented.

Total Productive Maintenance (TPM)

TPM has been a well-known activity with several names associated with it. Many people associate TPM with Total Predictive Maintenance or Total Preventative Maintenance. The association explained here will be Total Productive Maintenance but includes these other terms as well.

TABLE 11.11

Setup Conversion Matrix

SETUP CONVERSION MATRIX Sheet				Date:	Page	of
Area/Department	Machine/Equipment Name	Setup Tools Required		Operator Number	Standard Setup Time	
				Date Prepared	Minutes	
	CURRENT PROCESS	**CURRENT TIME**		**IMPROVEMENT**	**PROPOSED TIME**	
NO.	Task/Operation	Internal	External		Internal	External

Current Total

Improve Total

Conversation Methodology

Preparation of Setup Process

Combining Equipment Functionality

Standardized Jigs

TPM is performed during the improvement phase based on downtimes or efficiency losses. The associated downtimes can be planned or unplanned. The goal of TPM is to increase all operational equipment efficiencies to above 85% by eliminating any wasted time such as setup time (see SMED section), idle times, downtimes, start-up delays, and any quality losses.

TPM ensures minimal downtime but in turn also requires no defects. There are three basic steps for TPM that have several steps within each.

1. Analyze the current processes
   a. Calculate any costs associated with maintenance.
   b. Calculate overall equipment effectiveness (OEE) by finding the proportion of quality products produced at a given line speed.

2. Restore equipment to its original and high-operating states
   a. Inspect the machinery.
   b. Clean the machinery.
   c. Identify necessary repairs that need to be made to the machinery.
   d. Document defects.
   e. Create a scheduling mechanism for maintenance.
   f. Ensure that maintenance has repaired machinery and improvements are sustained.

3. Preventative maintenance to be carried out
   a. Create a schedule for maintenance with priorities, including high machinery defects, replacement parts, and any other pertinent information.
   b. Create stable operations—Complete a root cause analysis on high machinery defects and machinery that causes major downtime.
   c. Create a planning and communication system—Documentation of preventative maintenance activities should be accessible to all people so planning and prioritizing is completed.
   d. Create processes for continuous maintenance—Inspections should occur regularly and servicing for any machinery should be noted on a scheduled basis.
   e. Internal operations should be optimized—Any internal operations should be benchmarked with improvements from other areas to eliminate time spent on root cause analysis. When defects of machinery are not understood, it is important to put the machinery back to its original state to understand the root causes more efficiently. Time for exchanging or retrieving parts should also be minimized.

f. Continuous improvement on preventative maintenance—Train employees for early detection of problems and maintenance measures. Visual controls should be put in place for changeovers. 5S should take place to eliminate wasted time. The documentation should be communicated and plans should be given regularly. All aspects should be looked upon to see if Continuous Improvements can be made.

The key TPM indicators should show the following main issues:

- OEE
- Mean time between failures
- Mean time to repair

TPM is crucial to sustainability because it involves all the employees including high-level managers and creates planning for preventative maintenance so that issues are fixed before they become an error or defect. TPM is also a journey for educating and training the workforce to be familiar with machinery, parts, processes, and damage while being productive.

Design for Six Sigma (DFSS)

Design for Six Sigma is another process that is included in a phase called DMADV, which stands for the following:

Define
Measure
Analyze
Design
Verify

The difference between DMADV and DMAIC is the design and verification portions. DMAIC is process improvement driven, whereas DMADV is for designing new products or services. Design stands for the designing of new processes required, including implementation.

Verify stands for the results being verified and the performance of the design to be maintained.

The purpose of DFSS is very similar to the regular DMAIC cycle where it is a customer-driven design of processes with Six Sigma capabilities. DFSS does not only have to be manufacturing driven; the same methodologies can be used in service industries. The process is top down with flow down CTQs that match flow up capabilities. DFSS is quality based where predictions are made regarding first-pass quality. The quality measurements are driven through predictability in the early design phases. Process capabilities are utilized to make final design decisions.

Finally, process variances are monitored to verify that Six Sigma customer requirements are met.

The main tools utilized in DFSS are FMEAs, Quality Function Deployment (QFD), Design of Experiments (DOE), and simulations.

Quality Function Deployment (QFD)

Dr. Yoji Akao developed Quality Function Deployment (QFD) in 1966 in Japan. There was a combination of quality assurance and quality control that led to value engineering analyses. The methods for QFD are simply to utilize consumer demands into designing quality functions and methods to achieve quality into subsystems and specific elements of processes. The basis for QFD is to take customer requirements from the Voice of the Customer and relay them into engineering terms to develop products or services. Graphs and matrices are utilized for QFD. A house type matrix is compiled to ensure customer needs are being met into the transformation of the processes or services designed. The QFD house (Figure 11.64) is a simple matrix

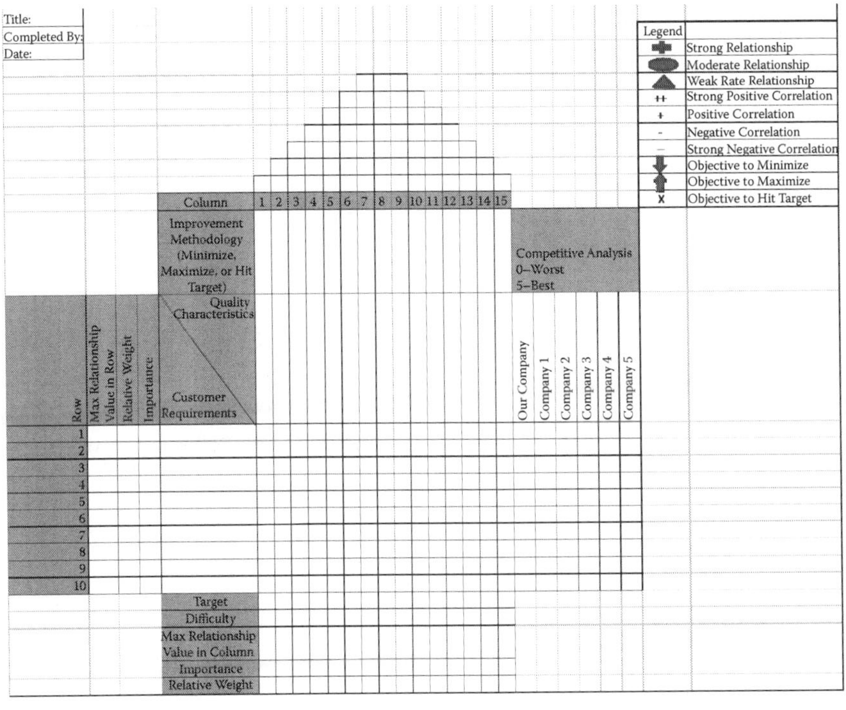

FIGURE 11.64
Quality Function Deployment House.

where the legend is used to understand quality characteristics, customer requirements, and completion.

Design of Experiments (DOE)

Design of Experiments (DOE) is an experimental design that shows what is useful, what has a negative connotation, and what has no effect. The majority of the time, 50% of the designs have no effect.

DOE requires a collection of data measurements, systematic manipulation of variables also known as factors placed in a prearranged way (experimental designs), and control for all other variables. The basis behind DOE is to test everything in a prearranged combination and measure the effects of each of the interactions.

The following DOE terms are used:

- *Factor*—An independent variable that may affect a response
- *Block*—A factor used to account for variables that the experimenter wishes to avoid or separate during analysis
- *Treatment*—Factor levels to be examined during experimentation
- *Levels*—Given treatment or setting for an input factor
- *Response*—The result of a single run of an experiment at a given setting (or given combination of settings when more than one factor is involved)
- *Replication (Replicate)*—Repeated run(s) of an experiment at a given setting (or given combination of settings when more than one factor is involved)

There are two types of DOE: Full-Factorial Design and Fractional Factorial Design. Full-Factorial DOE determine the effect of the main factors and factor interactions by testing every factorial combination. A Full-Factorial DOE factors all levels combined with one another covering all interactions. The basic design of a three-factorial DOE is shown in Table 11.12.

The effects from the Full-Factorial DOE can then be calculated and sorted into main effects and effects generated by interactions.

Effect = Mean value of response when factor setting is at high level (Y_A+) – Mean value of response when factor setting is at low level (Y_A-)

In a full factorial experiment, all of the possible combinations of factors and levels are created and tested.

TABLE 11.12

Full Factorial DOE

	Factors			Factor Interactions			
Number	A	B	C	AB	AC	BC	ABC
1	–	–	–	+	+	+	–
2		–	–	–	–	+	+
3	–	+	–	–	+	–	+
4	+	+	–	+	–	–	–
5	–	–	+	+	–	–	+
6	+	–	+	–	+	–	–
7	–	+	+	–	–	+	–
8	+	+	+	+	+	+	+

In a two-level design (where each factor has two levels) with k factors, there are 2^k possible scenarios or treatments.

- 2 factors each with 2 levels, we have $2^2 = 4$ treatments
- 3 factors each with 2 levels, we have $2^3 = 8$ treatments
- k factors each with 2 levels, we have 2^k treatments

The analysis behind the DOE consists of the following steps:

1. Analyze the data
2. Determine factors and interactions
3. Remove statistically insignificant effects from the model such as p-values of less than .1 and repeat the process
4. Analyze residuals to ensure the model is set correctly
5. Analyze the significant interactions and main effects on graphs while setting up a mathematical model
6. Translate the model into common solutions and make sustainable improvements

A Fractional Factorial Design locates the relationship between influencing factors in a process and any resulting processes while minimizing the number of experiments. Fractional Factorial DOE reduce the number of experiments while still ensuring the information lost is as minimal as possible. These types of DOE are used to minimize time spent, money spent, and eliminate factors that seem unimportant.

The formula for a Fractional Factorial DOE is $2^{k-q,}$ where q equals the reduction factor.

The Fractional Factorial DOE requires the same number of positive and negative signs as a Full-Factorial DOE.

The Fractional Factorial DOE is shown in a matrix in Table 11.13.

TABLE 11.13

Fractional Factorial DOE

Factor Factorial Matrix												D				
Run	**(I)**	**A**	**B**	**C**	**D**	**AB**	**AC**	**AD**	**BC**	**BD**	**CD**	**ABC**	**ABD**	**ACD**	**BCD**	**ABCD**
1	+	−	−	−	−	+	+	+	+	+	+	−	−	−	−	+
2	+	+	−	−	+	−	−	+	+	−	−	+	−	−	+	+
3	+	−	+	−	+	−	+	−	−	+	−	+	−	+	−	+
4	+	+	+	−	−	+	−	−	−	−	+	−	−	+	+	+
5	+	−	−	+	+	+	−	−	−	−	+	+	+	−	−	+
6	+	+	−	+	−	−	+	−	−	+	−	−	+	−	+	+
7	+	−	+	+	−	−	−	+	+	−	−	−	+	+	−	+
8	+	+	+	+	+	+	+	+	+	+	+	+	+	+	+	+

Mood's Median Test

Mood's Median Test compares the medians of different samples of data when nonnormal data is present and there are obvious outliers in the data.

Example 1: Mood's Median Test

Mood's Median Test: Temperature versus Location

```
Mood median test for Temperature
Chi-Square = 17.07 DF = 1 P = 0.000
                                   Individual 95.0% CIs
Location N< = N> Median Q3-Q1 — +— — — — -+— — — — -+— — — — -+— —
San Francisco 7 23 1.1282 0.0306                (— — *— — )
Cleveland 23 7 1.0882 0.0482 (— — -*— — -)
                              — +— — — — -+— — — — -+— — — — -+— —
                               1.080      1.100     1.120     1.140
Overall median = 1.1127
A 95.0% CI for median(San Francisco) - median(Cleveland): (0.0277,0.0523)
```

Based on the data in the subgroups, the difference between the subgroup medians is .04 (1.1282 – 1.0882). The confidence interval for the difference provides more detail and confirms (with 95% confidence) that the difference in medians is somewhere between .0277 and .0523. The p-value for the test is 0.0. Since this is less than an alpha level of .05 we can say, with 95% confidence, that the medians of the subgroups are different. A rough graph of the 95% confidence intervals (CIs) for the medians of the subgroups is shown above. The statistical conclusion is that the medians are different.

To summarize the results, we can be very confident that there is a difference in the median ratios for temperature between San Francisco and Cleveland.

Example 2: Mood's Median Test

Mood's Median Test: Water Solubility versus Location

```
Mood median test for Water Solubility
Chi-Square = 0.07 DF = 1 P = 0.796
                                   Individual 95.0% CIs
Location N< = N> Median Q3-Q1 — — — +— — — — -+— — — — -+— — — — -+
San Francisco 15 15 5.73 1.84 (-*— — — — — — — — — — — — -)
Cleveland 16 14 5.70 2.04 (*— — — — — — — — — — — — — — — — -)
                              — — — +— — — — -+— — — — -+— — — — -+
                                   6.00      6.60      7.20      7.80
Overall median = 5.70
A 95.0% CI for median(San Francisco) - median(Cleveland): (-1.82,1.43)
```

Based on the data in the subgroups, the difference between the subgroup medians is .03 (5.73 – 5.7). The confidence interval for the difference provides more detail and confirms (with 95% confidence) that the difference in medians is somewhere between –1.82 and 1.43. The p-value for the test is .796. Since this is more than an alpha level of .05 we can say, with 95% confidence, that the medians of the subgroups are not different. A rough graph of the 95% confidence intervals (CIs) for the medians of the subgroups is shown above with overlap. The statistical conclusion is that the medians are the same.

To summarize the results, we can be very confident that there is no difference in the median ratios for water solubility between San Francisco and Cleveland.

Control Plans

A control plan is a vital part of sustainability because without it, there is no sustainability. A control plan takes the improvements made and ensures that they are being maintained and Continuous Improvement is achieved. A control plan is a very detailed document that includes who, what, where, when, and why (the why is based on the root cause analysis). The 12 basic steps of a control plan are:

1. Collect existing documentation for the process
2. Determine the scope of the process for the current control plan
3. Form teams to update the control plan regularly
4. Replace short-term capability studies with long-term capability results
5. Complete control plan summaries
6. Identify missing or inadequate components or gaps
7. Review training, maintenance, and operational action plans
8. Assign tasks to team members
9. Verify compliance of actual procedures with documented procedures
10. Retrain operators
11. Collect sign-offs from all departments
12. Verify effectiveness with long-term capabilities

A control plan ensures consistency while eliminating as much variation from the system as possible. The plans are essential to operators because it enforces SOPs and eliminates changes in processes. It also ensures that preventative maintenance (PM) is performed and the changes made to the processes actually improve the problem that was found through the root

Six Sigma Control Plan											
Product:				Core Team:					Date: (orig):		
Key Contact:									Date: (revised):		
Phone:											
Process	Process Step	Input	Output	Process Specs (LSL, USL, Target)	Pkg./Date	Measurement Technique	%P/T	Sample Size	Sample Frequency	Control Method	Reaction Plan

FIGURE 11.65
Control Plan.

cause analysis. Control plans hold people accountable if reviewed at least quarterly.

A sample control plan is shown in Figure 11.65.

References

Agustiady, Tina, and Adedeji B. Badiru. 2012. *Sustainability: Utilizing Lean Six Sigma Techniques*. Boca Raton, FL: Taylor & Francis/CRC Press.

Fiorino, Donald P. 2004. Case Study: Utility Cost Reduction at a Large Manufacturing Facility. http://texasiof.ces.utexas.edu/PDF/Presentations/Nov4/CaseStudyFiorino.pdf.

Karliner, Joshua. 2005. Green Schools Initiative, The David Brower Center, "Little Green Schoolhouse" Report.

Leap, Local Energy Alliance Program. 2010. Charlottesville, VA. http://leap-va.org/energy-ed-center/.

Manufacturing Institute. n.d. Energy Efficiency Toolkit for Manufacturers: Eight Proven Ways to Reduce Your Costs. http://www.energy.ca.gov/process/pubs/toolkit.pdf.

The Story of Stuff. 2007. The Story of Stuff Project. http://www.storyofstuff.com/.

Water Use Efficiency Branch. 2010. Sacramento, CA. http://www.water.ca.gov/.

Zero Energy Commercial Buildings Consortium. 2008. Commercial Buildings Initiative.

12

A Lean Six Sigma Case Study

> If you want to prosper for a year, grow rice. If you want to prosper for a decade, plant trees. If you to prosper for a century, grow people.
>
> —A wise old farmer reflecting back on a life of toil in the soil

The following Lean Six Sigma case study will reflect a real-life manufacturing problem with Continuous Improvement and Lean Six Sigma tools to show how some of the tools are put into place in the real world.

Case Study: Process Improvement—Argo Rework Reduction

Tina Agustiady, Certified Six Sigma Master Black Belt

Executive Summary

Argo is a growing product for a main dog toy factory. The products are selling extremely well and sales have been rising for the past year. Over the past year, with the increase of sales there has also been an increase in complaints in problems with the toys. Customers feel that some toys fall apart far too quickly. Poor material quality of the product results in major variation in product due to materials, methods, machinery, and so on. Many toys were being reworked in the factory causing more time and money while some customers were complaining as well.

Problem Statement: Reduce rework for all toys by 15% by March 2014 and 25% by June 2014, and have less than 1% customer complaints.

Methodology: Training, Equipment, Line Balancing, Layout, Process, Data-Gathering Root Cause Analysis

Introduction

Tina Agustiady—Continuous Improvement Leader

Buddy Lee—Operator

Mike Thomas—Operator

Brian Crops—Production Engineer

Tamara Brown—Maintenance Coordinator

Nelly Curtis—Plant Manager and Project Champion

Leo Downs—Executive Sponsor

Gweneth Verns—Master Black Belt

The team was selected based on knowledge and expertise of the process, and was proficient and organized during the project.

Define

A project charter was mapped out to show responsible personnel, problem statements, goals, and timelines (see also Figure 12.1).

1. Project Charter Purpose

    The purpose of the Continuous Improvement project is to remove rework in a dog toy factory utilizing a structured approach and being able to benchmark the findings across the manufacturing industrial processes.

Goals	Objectives
• Reduce rework by 15% by March 2014 and 25% by June 2014 and have less than 1% in customer complaints	• Determine XLine process along with specifications of dog toys • Determine if target stuffing in toys are accurate or need to be revised • Train all personnel on XLine process and create manual • Determine proper PM's for XLine

FIGURE 12.1
Goals and objectives.

2. Project Executive Summary

   There is an average of 315 toys reworked per day (based on the past 30 days of data) due to various issues.

   Reduce rework for all toys by 15% by March 2014 and 25% by June 2014, and have less than 1% customer complaints.

3. Project Overview

   Business Justification consists of reducing rework issues and customer complaints associated on the XLine and increasing first-pass quality.

Project Charter Purpose

The purpose of the Continuous Improvement project is to pilot the Continuous Improvement foundation in a dog toy factory utilizing a structured approach and being able to benchmark the findings across manufacturing industrial processes (Figure 12.2).

A control map will be completed at the end of the project.

A process map was drawn out in order to visually see each of the process steps as they occur. The ANSI symbols were used to understand what types of processes were being accomplished (Figure 12.3).

Conclusion of Define: The Define stage showed that the processes committed for Argo are important to the customers and the business. A process map was completed in order to better fully understand the steps. The project team now has a baseline to begin the Measure phase through the process steps.

Milestone	Deliverable
1. Conduct training	• Training conducted March 2014
2. Create a manual	• The manual will be for documentation and training purposes, completion date August 2014
3. Create PMs for main equipment	• PMs will be established and sign-off sheets will be available, completion date September 2014
4. Benchmark best practice processes	• Determine Best in Class process and implement for XLine, completion date September 30, 2014
5. Reduce rework by 25% by June 2014	• Determine rework causes and reduce by 15% by March 2014 • Determine rework causes and reduce by 25% by June 2014

FIGURE 12.2
Milestones and deliverables.

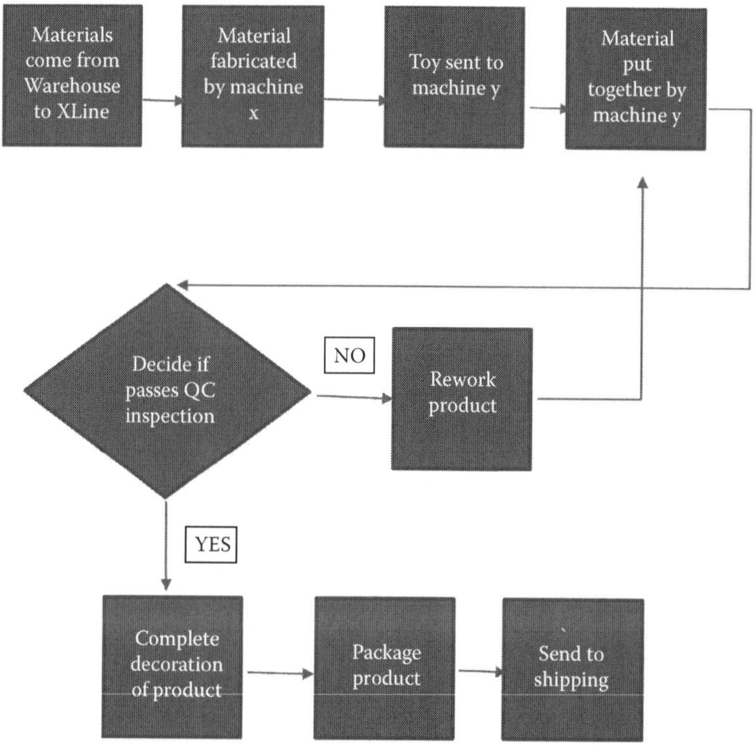

FIGURE 12.3
Process map.

Measure

The goals of the Measure phase were to determine the key factors for variation issues, machine issues, and variation in Argo processes through statistical analysis and graphical analysis. The goal of the project is to reduce rework by 15% by March 2014 and by 25% by June 2014 while having less than 1% customer complaints.

A SIPOC was performed to see the high-level process inputs, outputs, and process steps (Figure 12.4). A cause and effect diagram was created to ensure the importance of the different variables that play a part in the process. This was done for the XLine that had the most rework product (Figure 12.5). A cause and effect matrix was then completed to find the process' largest issues (Figure 12.6).

The next step was to perform a failure modes and effects analysis (FMEA) (Figure 12.7) after the cause and effect matrix. A Pareto chart was completed for the FMEA (Figure 12.8). Automatically it could be seen according to the FMEA that bad material or fabrication was an issue. The specifications were looked at as well to ensure that the product materials were within specification.

Suppliers	Inputs	Input Specification	Processes	Gap	Outputs	Customers
Raw Bone	Thread	1.0–1.1	Materials come from warehouse to XLine		Raw material	Wally
Pet Doo	Material	2.0–2.9	Material fabricated by machine X		Fabricated material	Tget
Cryer	Bone	4.1–4.9	Toy sent to machine Y		1/2 processed toy	Kroman
Whiner	Cushion	.8–1.0	Material put together by machine Y		Toy complete for inspection	Giant Store
Happy Pup	Squeaker	5.5–6.5	QC Process	**Manual Process**	Toy complete for inspection	Pet Peeps
	Rope	10.1–10.9	Rework		Toy complete for inspection	
	Treat	2.4–2.10	Decorate product		1/2 processed toy	
			Package		Fully processed toy	
			Ship		Toy ready for customer	

FIGURE 12.4
SIPOC.

Measurement	Man	Method	Materials	Machinery	Environment
Thread	Management	How the thread is sewn	Thread	Machine x	Humidity
Material	Training	Decoration technique	Material	Machine y	Temperature
Bone	Experience	Placement of finished product on belt	Bone	Packer	Standing surface
Cushion	Operator knowledge	Placement of product to machine x	Cushion	Conveyor belt	Housekeeping
Squeaker	Attention to detail	Placement of product to machine y	Squeaker	Decorator	Lighting
Rope	SOPs/ Visuals	Initial toy placement on belt	Rope	Mover	Equipment spacing
Treat	Attitudes of operators	Method of racking off toys	Treat	Placer	People spacing
Specific gravity	Long hours	Placement of material	Packaging	Sewer	
Temperature	Ownership	Packaging method		Drizzler	

FIGURE 12.5
Cause and effect diagram.

Conclusion of Measure: Data was taken of as many parameters as possible before changing any variables. It was found that the material was making a significant impact on the process and there needed to be analysis on the FMEA and cause and effect matrix.

The following was accomplished for the Measure phase:

Process Mapping
Data Gathering
Cause and Effect Diagram
Root Cause Analysis Setup
Failure Modes and Effects Analysis

Analyze

Since the material was automatically seen as an issue the specifications were looked at. It was noted that there were two material suppliers:

- Cryer
- Whiner

V Line Process	**Rating of Importance to Customer**	10	10	8	10	10	10	8	10	10	9	7	10	10	8	10		
	Process Step		1	2	3	4	5	6	7	8	9	10	11	12	13	14	15	
			Fabricate	Second Fabrication	Insert Middle	Threaded Product	Reworked Process	Decorated Product	Packaged	Shipping								Total
	Process Step	**Process Input**																
1	Materials come from warehouse to XLine	Material	9	0	0	0	9	9	9	9								432
2	Material fabricated by machine X	Material	9	0	0	0	9	9	9	9								432
3	Toy sent to machine Y	Middle Product	9	0	0	0	9	9	9	9								432
4	Material put together by machine Y	Middle Product	9	0	0	0	9	9	9	9								432
5	QC Process	Refabrication	3	0	0	0	3	0	0	0								60
6	Rework	Refabrication	3	0	0	0	3	0	0	0								60
7	Decorate product	Toy	1	0	0	0	3	0	0	0								40
8	Package	Toy	9	0	0	0	9	0	0	0								180
9	Ship	Toy	9	0	0	0	9	3	3	3								264
10																		0

FIGURE 12.6
Cause and effect matrix.

11																		0
12																		0
13																		0
14																		0
15																		0
16																		0
17																		0
18																		0
19																		0
20																		0
																		0
Total		610	0	0	0	630	390	312	390	0	0	0	0	0	0	0		
		Lower Spec																
		Target																
		Upper Spec																

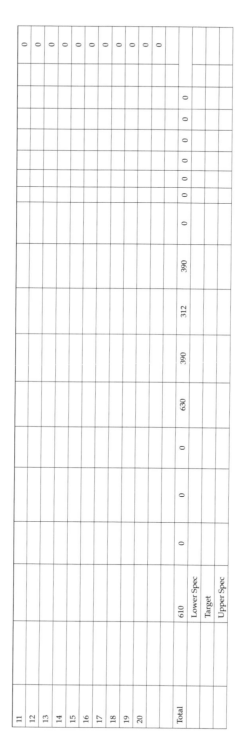

FIGURE 12.6
(***Continued***) Cause and effect matrix.

#	Process Function (Step)	Potential Failure Modes (process defects)	Potential Failure Effects (KPOVs)	SEV	Class	Potential Causes of Failure (KPIVs)	OCC	Current Process Controls		DET	RPN
1	Materials come from warehouse to XLine	Bad material	Torn product	8	XX	Supplier giving bad materials	9	QA Check	8	8	576
2	Material fabricated by machine X	Bad material or bad fabrication	Torn product	8	XX	Machine x problems	7	QA Check	10	10	560
3	Toy sent to machine Y	Bad material or bad fabrication	Dismantled product	8	XX	Machine x problems	8	QA Check	8	8	512
4	Material put together by machine Y	Bad material or bad fabrication	Dismantled product	7	X	Machine y problems	8	QA Check	8	8	448
5	QC Process	Manual check not completed correctly	Bad product sent out	8	XX	QC not complete	7	QA Check	7	7	392
6	Rework	Threading looks bad and customer complaint occurs	Bad product sent out	8	XX	QC not complete	7	QA Check	7	7	392
7	Decorate product	Decorating not properly completed	No or double decorating	8	XX	Machine does not notice already decorated	5	None	8	8	320
8	Package	Bad packaging	Bad product sent out	6	X	Packaging machine flaw	5	None	7	7	210
9	Ship	Shipping breaks product	Customer complaint for breakage	8	XX	Shipping flaw	6	None	2	2	96

FIGURE 12.7
Failure modes and effects analysis.

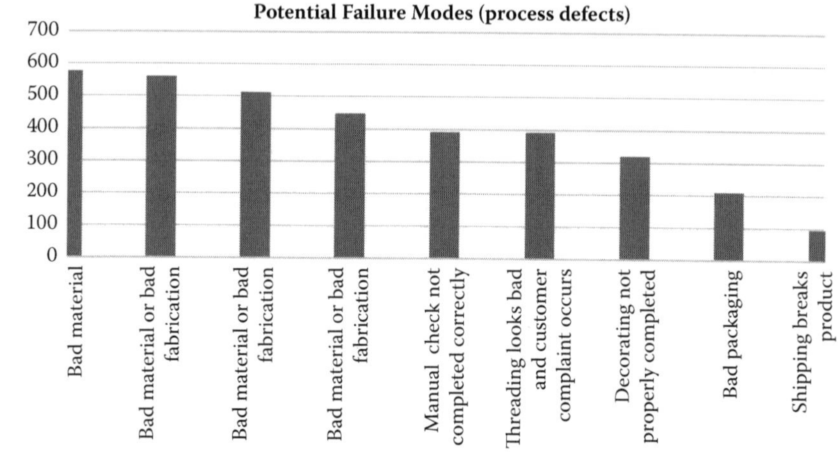

FIGURE 12.8
Failure modes and effects analysis Pareto chart.

The specifications were very different for the materials in terms of the thread count, which made a big difference in the material quality. The material needed to be between 2.0 and 2.9 (thousands). Cryer's specifications had an average of 2.6 (thousands). Whiner's specifications had an average of 2.3 (thousands). It was noted that Whiner's means were shifted to the left and needed to be centered (Figure 12.9).

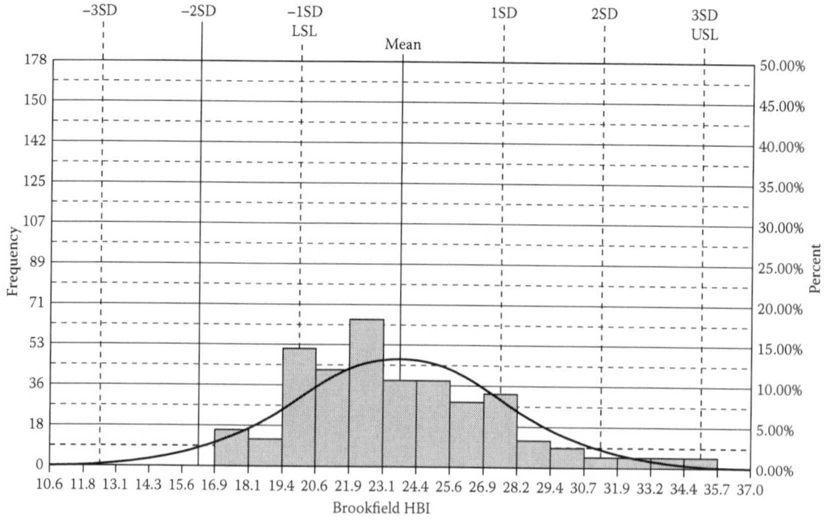

FIGURE 12.9
Means Shifted Analysis.

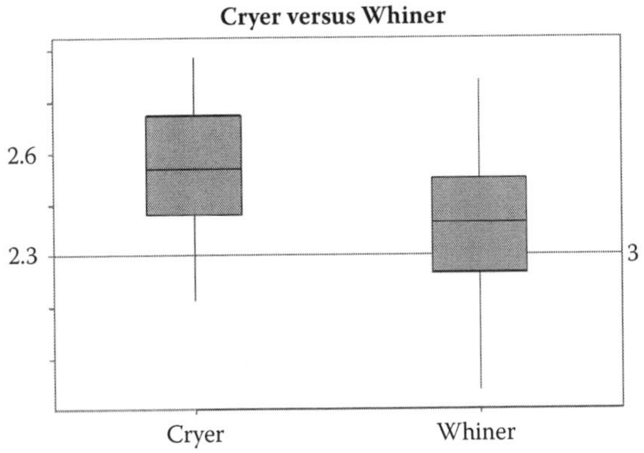

FIGURE 12.10
Boxplot for Cryer versus Whiner.

The boxplot concludes that Cryer has specifications targeted around the mean, whereas Whiner did not (Figure 12.10). The Certificates of Analysis (CofAs) were looked at to ensure that Whiner was giving in spec product. It was noted by viewing the CofAs that Whiner had a product shift in the past 2 months because it also had a supplier change. This was noted in the CofAs but not looked upon for the fabrication of Argo materials. Therefore, even though the supplier was noting that it had changed its supplier and had different specifications for Argo, the toy factory did not notice the changes in the CofA and continued to use the supplier.

The supplier needed to be talked to and told that the specifications were causing the factory some rework and customer complaints.

Conclusion of Analyze Phase: It was noted by viewing the CofAs that Whiner had a product shift in the past 2 months because they also had a supplier change. This was noted in the CofAs but not looked upon for the fabrication of Argo materials.

The supplier Whiner needed to be talked to and told that the specifications were causing the factory some rework and customer complaints.

Improve

The team decided to look upon the Cryer supplier to ensure it was giving good product. A graphical analysis was completed for Cryer (Figure 12.11). The graphical analysis proved to be what was needed for the factory to make the Argo product.

A hypothesis test was performed to see if Cryer and Whiner were performing the same, taking 30 random samples (Figure 12.12). The assumptions are shown in the null and alternate hypothesis:

H_0 = (the null hypothesis): The difference is equal to the chosen reference value $\mu1 - \mu2 = 0$

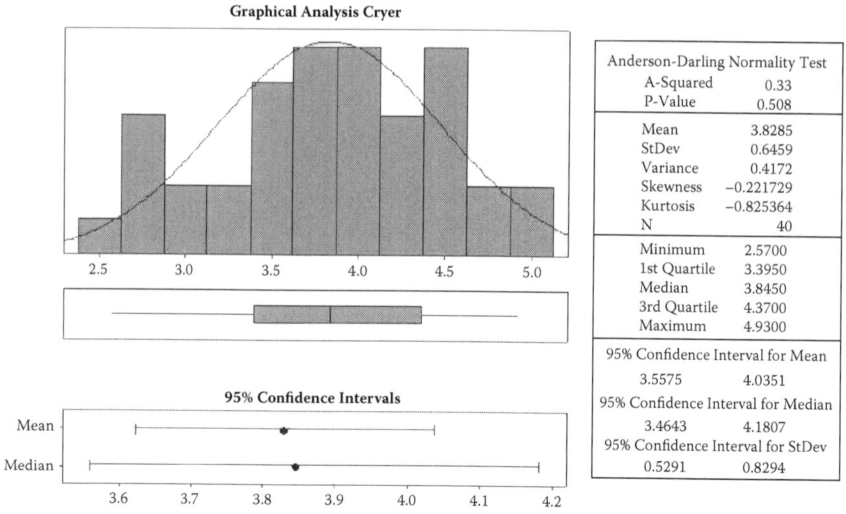

FIGURE 12.11
Graphical analysis for Cryer.

H_a = (The alternate hypothesis): The difference is not equal to the chosen reference value $\mu 1 - \mu 2 \neq 0$

95% CI for mean difference: (1.16, 6.69); t-test of mean difference = 0 (versus ≠ 0); t-value = 2.90; p-value = 0.007

The histogram of differences and the boxplot of differences are shown in Figures 12.13 and 12.14, respectively.

The hypothesis testing should be analyzed as follows: The confidence interval for the mean difference between the two materials does not include zero, which suggests a difference between them. The small p-value (p = 0.007) further suggests that the data are inconsistent with H_0 (null hypothesis) that is, the two materials do not perform equally. Specifically, the first set (mean = 79.697) performed better than the next set (mean = 83.623) in terms

Paired t-Test for Cryer versus Whiner				
	N	**Mean**	**Std Dev**	**SE Mean**
Cryer	30	82.62	5.2	0.95
Whiner	30	79.7	5	0.91
Difference	30	2.92	0.2	0.04

FIGURE 12.12
Paired t-test for Cryer versus Whiner.

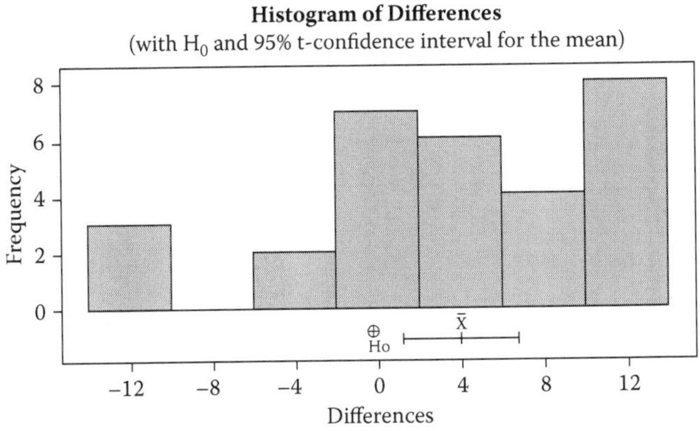

FIGURE 12.13
Histogram of differences t-test.

of differences between factories. Conclusion, reject H_0, the difference is not equal to the chosen reference value: $\mu1 - \mu2 \neq 0$.

The confidence interval for the mean difference between the two materials does not include zero, which suggests a difference between them. The small p-value ($p = 0.007$) further suggests that the data are inconsistent with H_0 (null hypothesis), that is, the two materials do not perform equally. Specifically, the first set (mean = 79.697) performed better than the other set (mean = 83.623) in terms of specifications. Conclusion, reject H_0, the difference is not equal to the chosen reference value: $\mu1- \mu2 \neq 0$.

Therefore, the materials seemed to be a major issue, and Cryer was performing better than Whiner. Therefore, the Whiner supplier needed to be let

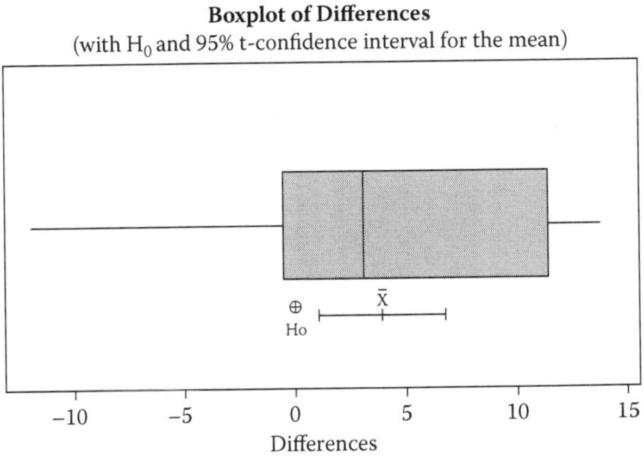

FIGURE 12.14
Boxplot of differences t-test.

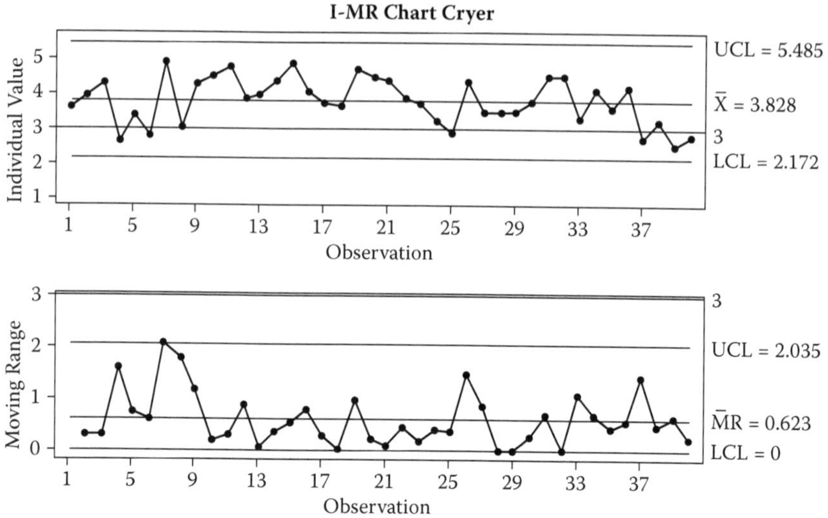

FIGURE 12.15
I-MR chart for Cryer.

go. It was seen that Whiner was not performing properly, causing material issues, rework, and customer complaints.

An I-MR chart was plotted for Cryer to ensure the specifications were performing as planned and giving good material (Figure 12.15). The I-MR Chart showed that all points were in control after taking the 30 data points for Cryer, which is a positive result that should be seen for root cause analysis.

The normality test (Figure 12.16) came out above .05 after looking at the 30 data points, which is what was sought after (.547 for Cryer). This proved again that Cryer was performing well.

A capability analysis was done on Cryer for the 30 random data points on the B side of the line (Figure 12.17).

Capability analyses were performed for Cryer. The Cpk (short term) value of 1.27 shows that the capability was desirable. The Ppk (long term) value of 0.94 could be improved, but was also desirable (a value of 1.33 is considered acceptable). The PPM Total (Exp. Overall Performance) is the number of parts per million (3203.17) whose characteristic of interest is outside the tolerance limits. This means that approximately 3203.17 out of 1 million batches do not meet the specifications. The improvements will be tracked again after the results with more sustained product samples.

Therefore, having a second supplier providing out-of-specification product (Whiner—not performing while Cryer—performing) was a root cause of the rework and customer complaint issues.

It was decided to trial just using supplier Cryer for a few months trial period and eliminating supplier Whiner.

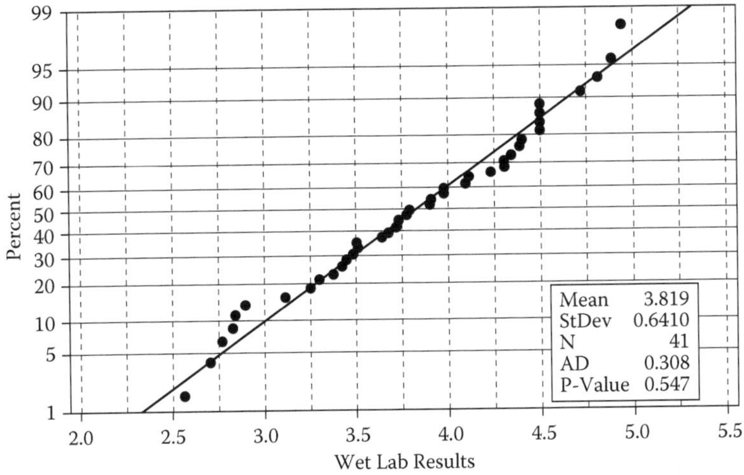

FIGURE 12.16
Normality chart for Cryer.

A Gage R&R was performed to ensure the supplier was able to have material with Repeatable and Reproducible product (Figures 12.18 and 12.19).
Gage R&R for Measurement A
Gage R&R Study—ANOVA Method
Two-Way ANOVA Table with Interaction

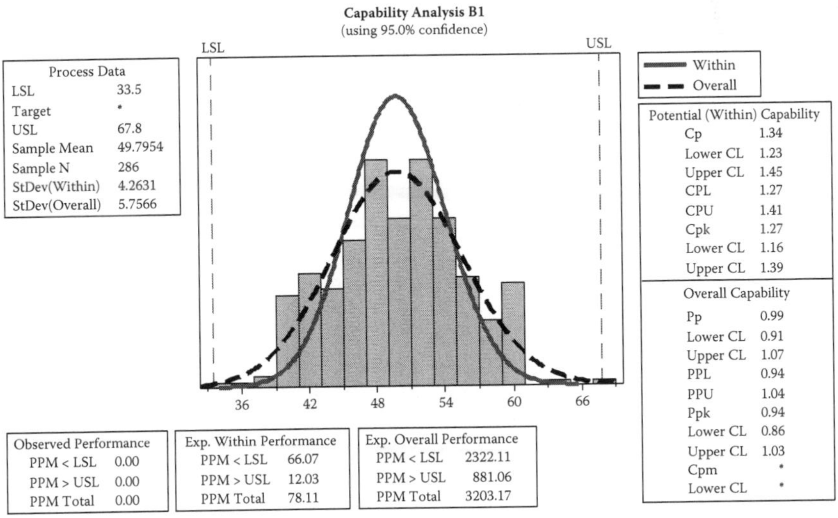

FIGURE 12.17
Capability analysis for Cryer.

Source	DF	SS	MS	F	P
Sample	9	39.9246	4.43606	308.960	0.000
Operator	2	0.0082	0.00411	0.286	0.754
Sample * Operator	18	0.2584	0.01436	1.700	0.065
Repeatability	60	0.5067	0.00844		
Total	89	40.6979			

Alpha to remove interaction term = 0.25

Gage R&R

Source	VarComp	%Contribution (of VarComp)
Total Gage R&R	0.010416	2.08
Repeatability	0.008444	1.68
Reproducibility	0.001971	0.39
Operator	0.000000	0.00
Operator*Sample	0.001971	0.39
Part-To-Part	0.491300	97.92
Total Variation	0.501716	100.00

Source	StdDev (SD)	Study Var (6 * SD)	%Study Var (%SV)
Total Gage R&R	0.102057	0.61234	14.41
Repeatability	0.091894	0.55136	12.97
Reproducibility	0.044398	0.26639	6.27
Operator	0.000000	0.00000	0.00
Operator*Sample	0.044398	0.26639	6.27
Part-To-Part	0.700928	4.20557	98.96
Total Variation	0.708319	4.24992	100.00

Number of Distinct Categories = 9

First, the interaction is checked for significance. The interaction is not significant unless the p-value is less than .05. In this case, the p-value is .065. Therefore, the interaction is not significant, which is what is desired from this study.

In the actual Gage R&R, the reproducibility is 12.97, which seems to be sufficient. The repeatability is 6.27. The variation should be reduced from the repeatability and reproducibility. If the repeatability is too high, the data gives just one distinct category. The standard deviation of the Gage R&R for part-to-part is .700. Again, the minimization of noise is desired.

The study variation in this case is .61, which is less than the allowable amount of 30. The P/T ratio is 14.41, which is close to the 15 or less that is ideal. The Gage is consuming a portion of 14% of the total variation in the measurement system. The Gage Run Study showed that Operator 3 had an inconsistent time keeping the same measurements from time to time.

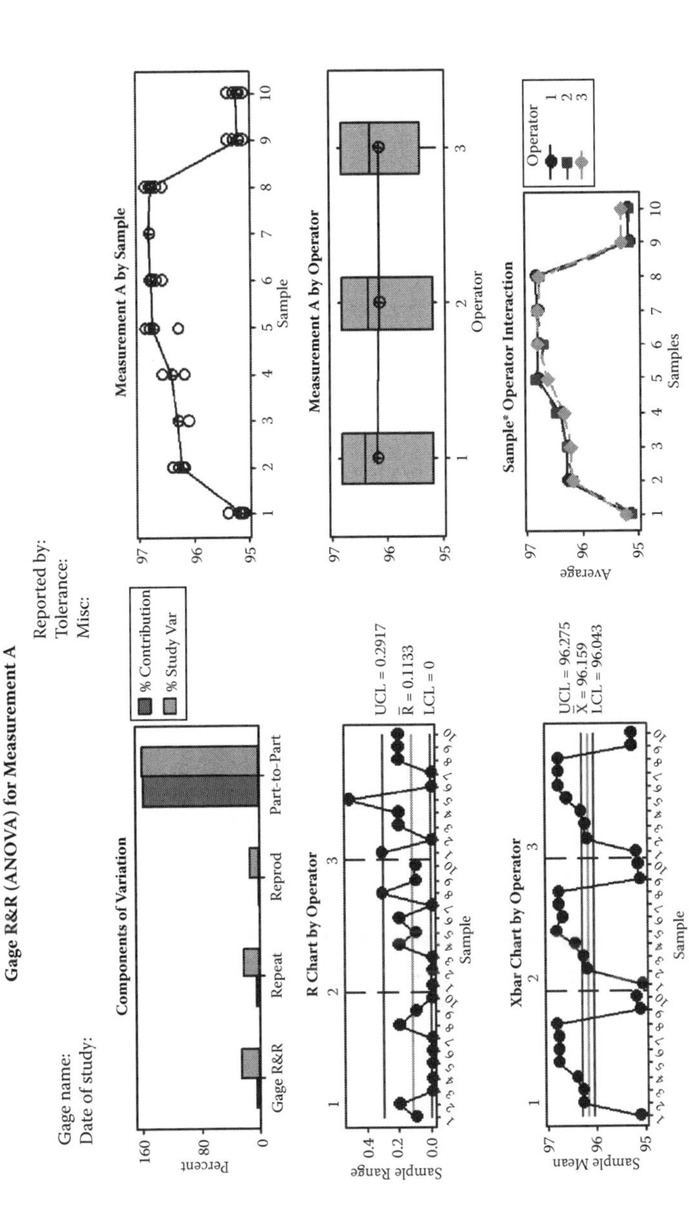

FIGURE 12.18
Gage R&R for Cryer.

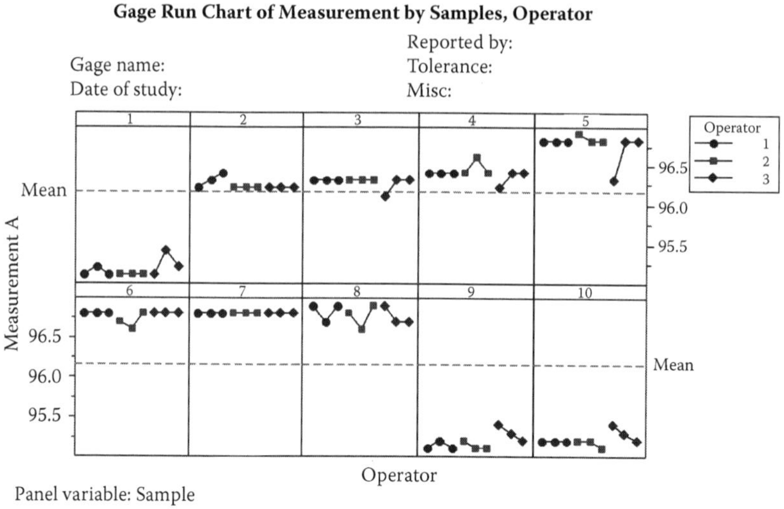

FIGURE 12.19
Gage R&R run chart.

The components of variation chart shows that there is low noise to high signal. The Gage R&R is the noise variable, and the repeatability plus the reproducibility make up the Gage R&R. The part-to-part is the signal. The number of distinct categories was 9. The Measurement System Analysis (MSA) was said to be a good study based on the results shown in Figure 12.20.

Conclusion of Improvement: Root cause analysis led to the evaluation of suppliers Cryer and Whiner. It was seen that Whiner was not performing the proper specifications, whereas Cryer was. It was trialed to only use Cryer and eliminate Whiner. Automatically a change in rework was seen where almost all rework was eliminated. There was a rework percentage of 5% total for Argo after the changes. The customer complaints were also reduced to 0.05% in the time period studied (6 months).

Control

The control plan in Figure 12.21 was completed once improvements were completed and sustained. An audit checklist (Figure 12.22) was completed to

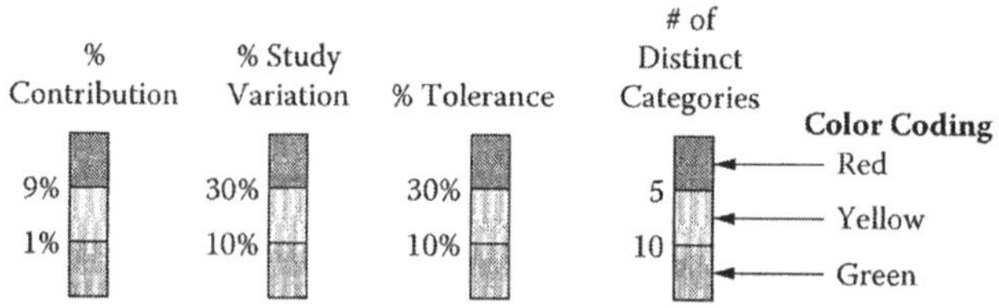

FIGURE 12.20
Good study results for Gage R&R.

	Control Plan									
	Process and Equipment Improvements	Prelaunch: YES	Production: YES	Key Contact/ Phone: Tina Agustiady	Tina Agustiady: 555-555-5555		Date (Orig.):	1-Oct-14		
Control Plan Number: 1A										
Part Number/ Latest Change Level				Core Team: Argo XLine			Customer Engineering Approval/ Date (If Req'd): Keith Smith 919-999-9999			
	Action Plans									
	Problem	**Action**	**Required Steps**	**Potential Results**	**Costs**	**Person Accountable**	**Due Date**	**Sample Size**	**Frequency**	**Control Method**
1	Supplier not giving in spec product	Check CofAs for specs and ensure in spec	Check each shipment	Ensure accuracy of material and will ensure accuracy	$ -	Keith Smith	11/1/2014	1	1x/month	Check and record specs on provided sheets
2	Specifications are only within 1 standard deviation	Make product at high and low end at 2 standard deviations for viscosity and show customer for acceptance	R&D and Marketing input and buy off	Increase of specifications for thread count	$ 10,000	John Clese	11/1/2014			Complete standard deviation and spec changes. Provide customer 2 samples per month. Change specifications with customer buy off

FIGURE 12.21
Control plan. (***Continued***)

3	Existing equipment in Facility B have experienced significant wear and are outdated		Purchase shafts, replace tube and bearing	1. Ensures product can be properly scraped off heat transfer tube wall	$ 36,200	Operations, David Johns, Kevin Tooks	11/1/2014			Check votators monthly for issues, preventative maintenance for votators to be completed 1x/month
	Replace shaft on equipment #1 with welded pins and 24″ shaft Replace heat transfer tube, scraper blades, and roller bearing in equipment #2			2. Reduces product build up resulting in decreased cleanouts, decreased growth of crystal structures and further wear and/or damage to the equipment 3. Requires replacement of shaft blades every 30 days along with pins creating extensive downtime						

FIGURE 12.21
Control plan.

4	Blades break in equipment	Replace equipment blades and put on PM for every 3 weeks	Put blades on PMs	Reduction in blades breaking and increase in preventative maintenance	$ 201	Operations, David Johns, Kevin Tooks	11/1/2014		1x/batch	Preventative maintenance to be checked daily, sign offs to be completed and Operations to be held accountable
5	Broad or no specifications for certain input variables	Proposal for input specifications	Production to ensure specifications are being followed	Reduction in variation		Operations, David Johns, Kevin Tooks	11/1/2014	Every batch	1x/batch	Check and record specs on provided sheets

FIGURE 12.21
Control plan.

Audit Technique	Auditable Item, Observation, Procedure, etc.	Individual Auditor Rating (Circle Rating)	
Observation	Have all associates been trained?	YES	NO
Observation	Is training documentation available?	YES	NO
Observation	Is training documentation current?	YES	NO
Observation	Are associates wearing proper safety gear?	YES	NO
Observation	Are SOPs available?	YES	NO
Observation	Are SOPs current?	YES	NO
Observation	Is quality being measured?	YES	NO
Observation	Is sampling being its sample size target?	YES	NO
Observation	Is the sampling meeting its sample size target?	YES	NO
Observation	Are control charts in control?	YES	NO
Observation	Are control charts current?	YES	NO
Observation	Is the process capability index >1.0?	YES	NO
Number of Out of Compliance Observations			
Total Observations			
Audit Yield			#DIV/0!
Corrective Actions Required			
Auditor Comments			

FIGURE 12.22
Audit checklist.

ensure that the problem would not to happen again in the future. The audit checklist will continually be audited to ensure the manufacturing groups are on track. The latest audit shows all items are on track.

Conclusion to Project: Many Lean Six Sigma tools were completed for root cause analysis for Product Argo in order to reduce rework and customer complaints. The root cause seemed to be due to the supplier Whiner who was eliminated from the factory. Many statistics were taken after the change to ensure the one supplier was up to specifications and producing no rework or customer complaints. The results are positive and will continue to be monitored.

Reference

Agustiady, Tina, and Adedeji B. Badiru. 2012. *Sustainability: Utilizing Lean Six Sigma Techniques*. Boca Raton, FL: Taylor & Francis/CRC Press.

Appendix

Useful Statistical Distributions

Industrial engineering uses statistical distributions and methods extensively in design and process improvement applications. The most common distributions are summarized in Table A.1.

Discrete Distributions

Probability Mass Function, $p(x)$

Mean, μ

Variance, σ^2

Coefficient of Skewness, β_1

Coefficient of Kurtosis, β_2

Moment-generating Function, $M(t)$

Characteristic Function, $\phi(t)$

Probability-Generating Function, $P(t)$

Bernoulli Distribution

$$p(x) = p^x q^{x-1} \qquad x = 0, 1 \qquad 0 \le p \le 1 \qquad q = 1 - p$$

$$\mu = p \qquad \sigma^2 = pq \qquad \beta_1 = \frac{1-2p}{\sqrt{pq}} \qquad \beta_2 = 3 + \frac{1-6pq}{pq}$$

$$M(t) = q + pe^t \qquad \phi(t) = q + pe^{it} \qquad P(t) = q + pt$$

TABLE A.1

Summary of Common Statistical Distributions

Distribution of Random Variable x	Functional Form	Parameters	Mean	Variance	Range
Binomial	$P_x(k) = \frac{n!}{k!(n-k)!} p^k (1-p)^{n-k}$	n, p	np	$np(1-p)$	$0, 1, 2, \ldots, n$
Poisson	$P_x(k) = \frac{\lambda^k e^{-\lambda}}{k!}$	λ	λ	λ	$0, 1, 2, \ldots$
Geometric	$P_x(k) = p(1-p)^{k-1}$	p	$1/p$	$\frac{1-p}{p^2}$	$1, 2, \ldots$
Exponential	$f_x(y) = \frac{1}{\theta} e^{-y/\theta}$	θ	θ	θ^2	$(0, \infty)$
Gamma	$f_x(y) = \frac{1}{\Gamma(\alpha)\beta^\alpha} y^{(\alpha-1)} e^{-y/\beta}$	α, β	$\alpha\beta$	$\alpha\beta^2$	$(0, \infty)$
Beta	$f_x(y) = \frac{\Gamma(\alpha+\beta)}{\Gamma(\alpha)\Gamma(\beta)} y^{(\alpha-1)} (1-y)^{(\beta-1)}$	α, β	$\frac{\alpha}{\alpha+\beta}$	$\frac{\alpha\beta}{(\alpha+\beta)^2(\alpha+\beta+1)}$	$(0, 1)$
Normal	$f_x(y) = \frac{1}{\sqrt{2\pi}\sigma} e^{-(y-\mu)^2/2\sigma^2}$	μ, σ	μ	σ^2	$(-\infty, \infty)$
Student t	$f_x(y) = \frac{1}{\sqrt{\pi v}} \frac{\Gamma\left(\frac{v+1}{2}\right)}{\Gamma(v/2)} (1 + y^2/v)^{-(v+1)/2}$	ν	0 for $\nu > 1$	$\frac{v}{v-2}$ for $v>2$	$(-\infty, \infty)$
Chi Square	$f_x(y) = \frac{1}{2^{v/2}\Gamma(v/2)} y^{(v-2)/2} e^{-y/2}$	ν	ν	2ν	$(0, \infty)$
F	$f_x(y) = \frac{\Gamma\left(\frac{v_1+v_2}{2}\right) v_1^{v_1/2} v_2^{v_2/2}}{\Gamma\left(\frac{v_1}{2}\right)\Gamma\left(\frac{v_2}{2}\right)} \frac{(y)^{(v_1/2)-1}}{(v_2+v_1 y)^{\frac{v_1+v_2}{2}}}$	v_1, v_2	$\frac{v_2}{v_2-2}$ for $\nu_2 > 2$	$\frac{v_2^2(2v_2+2v_1-4)}{v_1(v_2-2)^2(v_2-4)}$ for $\nu_2 > 4$	$(0, \infty)$

Beta Binomial Distribution

$$p(x)=\frac{1}{n+1}\frac{B(a+x,b+n-x)}{B(x+1,n-x+1)B(a,b)} \qquad x=0,1,2,\ldots,n \qquad a>0 \qquad b>0$$

$$\mu=\frac{na}{a+b} \qquad \sigma^2=\frac{nab(a+b+n)}{(a+b)^2(a+b+1)}$$

$B(a,b)$ is the Beta function.

Beta Pascal Distribution

$$p(x)=\frac{\Gamma(x)\Gamma(\nu)\Gamma(\rho+\nu)\Gamma(\nu+x-(\rho+r))}{\Gamma(r)\Gamma(x-r+1)\Gamma(\rho)\Gamma(\nu-\rho)\Gamma(\nu+x)} \qquad x=r,r+1,\ldots \quad \nu>p>0$$

$$\mu=r\frac{\nu-1}{\rho-1},\ \rho>1 \qquad \sigma^2=r(r+\rho-1)\frac{(\nu-1)(\nu-\rho)}{(\rho-1)^2(\rho-2)},\ \rho>2$$

Binomial Distribution

$$p(x)=\binom{n}{x}p^x q^{n-x} \quad x=0,1,2,\ldots,n \quad 0\le p\le 1 \quad q=1-p$$

$$\mu=np \qquad \sigma^2=npq \qquad \beta_1=\frac{1-2p}{\sqrt{npq}} \qquad \beta_2=3+\frac{1-6pq}{npq}$$

$$M(t)=\left(q+pe^t\right)^n \qquad \phi(t)=\left(q+pe^{it}\right)^n \qquad P(t)=\left(q+pt\right)^n$$

Discrete Weibull Distribution

$$p(x)=(1-p)^{x^\beta}-(1-p)^{(x+1)^\beta} \qquad x=0,1,\ldots \quad 0\le p\le 1 \quad \beta>0$$

Geometric Distribution

$$p(x)=pq^{1-x} \quad x=0,1,2,\ldots \quad 0\le p\le 1 \quad q=1-p$$

$$\mu=\frac{1}{p} \qquad \sigma^2=\frac{q}{p^2} \qquad \beta_1=\frac{2-p}{\sqrt{q}} \qquad \beta_2=\frac{p^2+6q}{q}$$

$$M(t)=\frac{p}{1-qe^t} \qquad \phi(t)=\frac{p}{1-qe^{it}} \qquad P(t)=\frac{p}{1-qt}$$

Hypergeometric Distribution

$$p(x)=\frac{\binom{M}{x}\binom{N-M}{n-x}}{\binom{N}{n}} \qquad x=0,1,2,\ldots,n \quad x\le M \quad n-x\le N-M$$

$$n, M, N, \in N; \quad 1\le n\le N; 1\le M\le N; \quad N=1,2,\ldots$$

$$\mu=n\frac{M}{N} \quad \sigma^2=\left(\frac{N-n}{N-1}\right)n\frac{M}{N}\left(1-\frac{M}{N}\right) \quad \beta_1=\frac{(N-2M)(N-2n)\sqrt{N-1}}{(N-2)\sqrt{nM(N-M)(N-n)}}$$

$$\beta_2=\frac{N^2(N-1)}{(N-2)(N-3)nM(N-M)(N-n)}$$

$$\left\{N(N+1)-6n(N-n)+3\frac{M}{N^2}(N-M)\left[N^2(n-2)-Nn^2+6n(N-n)\right]\right\}$$

$$M(t)=\frac{(N-M)!(N-n)!}{N!}F\left(.,e^t\right)$$

$$\phi(t)=\frac{(N-M)!(N-n)!}{N!}F\left(.,e^{it}\right) \quad P(t)=\left(\frac{N-M}{N}\right)^n F(.,t)$$

$F(\alpha,\beta,\gamma,x)$ is the hypergeometric function. $\alpha=-n; \beta=-M; \gamma=N-M-n+1$

Negative Binomial Distribution

$$p(x)=\binom{x+r-1}{r-1}p^r q^x \quad x=0,1,2,\ldots \quad r=1,2,\ldots \quad 0\le p\le 1 \quad q=1-p$$

$$\mu=\frac{rq}{p} \quad \sigma^2=\frac{rq}{p^2} \quad \beta_1=\frac{2-p}{\sqrt{rq}} \quad \beta_2=3+\frac{p^2+6q}{rq}$$

$$M(t)=\left(\frac{p}{1-qe^t}\right)^r \quad \phi(t)=\left(\frac{p}{1-qe^{it}}\right)^r \quad P(t)=\left(\frac{p}{1-qt}\right)^r$$

Poisson Distribution

$$p(x)=\frac{e^{-\mu}\mu^x}{x!} \qquad x=0,1,2,\ldots \qquad \mu>0$$

$$\mu = \mu \qquad \sigma^2 = \mu \qquad \beta_1 = \frac{1}{\sqrt{\mu}} \qquad \beta_2 = 3 + \frac{1}{\mu}$$

$$M(t) = \exp\left[\mu\left(e^t - 1\right)\right] \qquad \sigma(t) = \exp\left[\mu\left(e^{it} - 1\right)\right] \qquad P(t) = \exp[\mu(t-1)]$$

Rectangular (Discrete Uniform) Distribution

$$p(x) = 1/n \qquad x = 1, 2, \ldots, n \qquad n \in N$$

$$\mu = \frac{n+1}{2} \qquad \sigma^2 = \frac{n^2 - 1}{12} \qquad \beta_1 = 0 \qquad \beta_2 = \frac{3}{5}\left(3 - \frac{4}{n^2 - 1}\right)$$

$$M(t) = \frac{e^t\left(1 - e^{nt}\right)}{n\left(1 - e^t\right)} \qquad \phi(t) = \frac{e^{it}\left(1 - e^{nit}\right)}{n\left(1 - e^{it}\right)} \qquad P(t) = \frac{t\left(1 - t^n\right)}{n(1 - t)}$$

Continuous Distributions

Probability Density Function, $f(x)$

Mean, μ

Variance, σ^2

Coefficient of Skewness, β_1

Coefficient of Kurtosis, β_2

Moment-Generating Function, $M(t)$

Characteristic Function, $\phi(t)$

Arcsin Distribution

$$f(x) = \frac{1}{\pi\sqrt{x(1-x)}} \qquad 0 < x < 1$$

$$\mu = \frac{1}{2} \qquad \sigma^2 = \frac{1}{8} \qquad \beta_1 = 0 \qquad \beta_2 \frac{3}{2}$$

Beta Distribution

$$f(x) = \frac{\Gamma(\alpha + \beta)}{\Gamma(\alpha)\Gamma(\beta)} x^{\alpha - 1}(1 - x)^{\beta - 1} \qquad 0 < x < 1 \qquad \alpha, \beta > 0$$

$$\mu = \frac{\alpha}{\alpha+\beta} \quad \sigma^2 = \frac{\alpha\beta}{(\alpha+\beta)^2(\alpha+\beta+1)} \quad \beta_1 = \frac{2(\beta-\alpha)\sqrt{\alpha+\beta+1}}{\sqrt{\alpha\beta}(\alpha+\beta+2)}$$

$$\beta_2 = \frac{3(\alpha+\beta+1)\left[2(\alpha+\beta)^2+\alpha\beta(\alpha+\beta-6)\right]}{\alpha\beta(\alpha+\beta+2)(\alpha+\beta+3)}$$

Cauchy Distribution

$$f(x) = \frac{1}{b\pi\left(1+\left(\frac{x-a}{b}\right)^2\right)} \qquad -\infty < x < \infty \qquad -\infty < a < \infty \qquad b > 0$$

$\mu, \sigma^2, \beta_1, \beta_2, M(t)$ do not exist. $\phi(t) = \exp[ait - b|t|]$

Chi Distribution

$$f(x) = \frac{x^{n-1}e^{-x^2/2}}{2^{(n/2)-1}\Gamma(n/2)} \qquad x \geq 0 \qquad n \in N$$

$$\mu = \frac{\Gamma\left(\frac{n+1}{2}\right)}{\Gamma\left(\frac{n}{2}\right)} \qquad \sigma^2 = \frac{\Gamma\left(\frac{n+2}{2}\right)}{\Gamma\left(\frac{n}{2}\right)} - \left[\frac{\Gamma\left(\frac{n+1}{2}\right)}{\Gamma\left(\frac{n}{2}\right)}\right]^2$$

Chi-Square Distribution

$$f(x) = \frac{e^{-x/2}x^{(\nu/2)-1}}{2^{\nu/2}\Gamma(\nu/2)} \qquad x \geq 0 \qquad \nu \in N$$

$$\mu = \nu \quad \sigma^2 = 2\nu \quad \beta_1 = 2\sqrt{2/\nu} \quad \beta_2 = 3 + \frac{12}{\nu} \quad M(t) = (1-2t)^{-\nu/2},\ t < \frac{1}{2}$$

$$\phi(t) = (1-2it)^{-\nu/2}$$

Erlang Distribution

$$f(x) = \frac{1}{\beta^{n(n-1)!}} x^{n-1}e^{-x/\beta} \qquad x \geq 0 \qquad \beta > 0 \qquad n \in N$$

$$\mu = n\beta \qquad \sigma^2 = n\beta^2 \qquad \beta_1 = \frac{2}{\sqrt{n}} \qquad \beta_2 = 3 + \frac{6}{n}$$

$$M(t) = (1 - \beta t)^{-n} \qquad \phi(t) = (1 - \beta it)^{-n}$$

Exponential Distribution

$$f(x) = \lambda e^{-\lambda x} \qquad x \geq 0 \qquad \lambda > 0$$

$$\mu = \frac{1}{\lambda} \qquad \sigma^2 = \frac{1}{\lambda^2} \qquad \beta_1 = 2 \qquad \beta_2 = 9 \qquad M(t) = \frac{\lambda}{\lambda - t}$$

$$\phi(t) = \frac{\lambda}{\lambda - it}$$

Extreme-Value Distribution

$$f(x) = \exp\left[-e^{-(x-\alpha)/\beta}\right] \qquad -\infty < x < \infty \qquad -\infty < \alpha < \infty \qquad \beta > 0$$

$\mu = \alpha + \gamma\beta$, $\gamma \doteq .5772$ … is Euler's constant $\sigma^2 = \frac{\pi^2\beta^2}{6}$

$$\beta_1 = 1.29857 \qquad \beta_2 = 5.4$$

$$M(t) = e^{\alpha t}\Gamma(1 - \beta t), \quad t < \frac{1}{\beta} \qquad \phi(t) = e^{\alpha it}\Gamma(1 - \beta it)$$

F Distribution

$$f(x) \frac{\Gamma\left[(\nu_1 + \nu_2)/2\right]\nu_1^{\nu_1/2}\nu_2^{\nu_2/2}}{\Gamma(\nu_1/2)\Gamma(\nu_2/2)} x^{(\nu_1/2)-1}(\nu_2 + \nu_1 x)^{-(\nu_1+\nu_2)/2}$$

$$x > 0 \qquad \nu_1, \nu_2 \in N$$

$$\mu = \frac{\nu_2}{\nu_2 - 2}, \nu_2 \geq 3 \qquad \sigma^2 = \frac{2\nu_2^2(\nu_1 + \nu_2 - 2)}{\nu_1(\nu_2 - 2)^2(\nu_2 - 4)}, \quad \nu_2 \geq 5$$

$$\beta_1 = \frac{(2\nu_1 + \nu_2 - 2)\sqrt{8(\nu_2 - 4)}}{\sqrt{\nu_1}(\nu_2 - 6)\sqrt{\nu_1 + \nu_2 - 2}}, \quad \nu_2 \geq 7$$

$$\beta_2 = 3 + \frac{12\left[(\nu_2-2)^2(\nu_2-4)+\nu_1(\nu_1+\nu_2-2)(5\nu_2-22)\right]}{\nu_1(\nu_2-6)(\nu_2-8)(\nu_1+\nu_2-2)}, \quad \nu_2 \geq 9$$

$$M(t) \text{ does not exist. } \phi\left(\frac{\nu_1}{\nu_2}t\right) = \frac{G(\nu_1,\nu_2,t)}{B(\nu_1/2,\nu_2/2)}$$

$B(a,b)$ is the Beta function. G is defined by

$$(m+n-2)G(m,n,t) = (m-2)G(m-2,n,t) + 2itG(m,n-2,t), \quad m,n>2$$

$$mG(m,n,t) = (n-2)G(m+2,n-2,t) - 2itG(m+2,n-4,t), \quad n>4$$

$$nG(2,n,t) = 2 + 2itG(2,n-2,t), \quad n>2$$

Gamma Distribution

$$f(x) = \frac{1}{\beta^\alpha \Gamma(\alpha)} x^{\alpha-1} e^{-x/\beta} \qquad x \geq 0 \qquad \alpha,\beta > 0$$

$$\mu = \alpha\beta \qquad \sigma^2 = \alpha\beta^2 \qquad \beta_1 = \frac{2}{\sqrt{\alpha}} \qquad \beta_2 = 3\left(1+\frac{2}{\alpha}\right)$$

$$M(t) = (1-\beta t)^{-\alpha} \qquad \phi(t) = (1-\beta it)^{-\alpha}$$

Half-Normal Distribution

$$f(x) = \frac{2\theta}{\pi}\exp\left[-\left(\theta^2 x^2/\pi\right)\right] \qquad x \geq 0 \qquad \theta > 0$$

$$\mu = \frac{1}{\theta} \qquad \sigma^2 = \left(\frac{\pi-2}{2}\right)\frac{1}{\theta^2} \qquad \beta_1 = \frac{4-\pi}{\theta^3} \qquad \beta_2 = \frac{3\pi^2-4\pi-12}{4\theta^4}$$

LaPlace (Double Exponential) Distribution

$$f(x) = \frac{1}{2\beta}\exp\left[-\frac{|x-\alpha|}{\beta}\right] \qquad -\infty < x < \infty \qquad -\infty < \alpha < \infty \qquad \beta > 0$$

$$\mu = \alpha \qquad \sigma^2 = 2\beta^2 \qquad \beta_1 = 0 \qquad \beta_2 = 6$$

$$M(t) = \frac{e^{\alpha t}}{1-\beta^2 t^2} \qquad \phi(t) = \frac{e^{\alpha it}}{1+\beta^2 t^2}$$

Logistic Distribution

$$f(x) = \frac{\exp[(x-\alpha)/\beta]}{\beta(1+\exp[(x-\alpha)/\beta])^2}$$

$$-\infty < x < \infty; \quad -\infty < \alpha < \infty; \quad -\infty < \beta < \infty$$

$$\mu = \alpha \qquad \sigma^2 = \frac{\beta^2\pi^2}{3} \qquad \beta_1 = 0 \qquad \beta_2 = 4.2$$

$$M(t) = e^{\alpha t}\pi\beta t \csc(\pi\beta t) \qquad \phi(t) = e^{\alpha it}\pi\beta it \csc(\pi\beta it)$$

Lognormal Distribution

$$f(x) = \frac{1}{\sqrt{2\pi}\sigma x}\exp\left[-\frac{1}{2\sigma^2}(1nx-\mu)^2\right]$$

$$x > 0; -\infty < \mu < \infty; \sigma > 0$$

$$\mu = e^{\mu+\sigma^2/2} \qquad \sigma^2 = e^{2\mu+\sigma^2}\left(e^{\sigma^2}-1\right)$$

$$\beta_1 = \left(e^{\sigma^2}+2\right)\left(e^{\sigma^2}-1\right)^{1/2} \qquad \beta_2 = \left(e^{\sigma^2}\right)^4 + 2\left(e^{\sigma^2}\right)^3 + 3\left(e^{\sigma^2}\right)^2 - 3$$

Noncentral Chi-Square Distribution

$$f(x) = \frac{\exp\left[-\frac{1}{2}(x+\lambda)\right]}{2^{\nu/2}}\sum_{j=0}^{\infty}\frac{x^{(\nu/2)+j-1\lambda j}}{\Gamma\left(\frac{\nu}{2}+j\right)2^{2j}j!}$$

$$x > 0; \lambda > 0; \nu \in N$$

$$\mu = \nu+\lambda \qquad \sigma^2 = 2(\nu+2\lambda) \qquad \beta_1 = \frac{\sqrt{8}(\nu+3\lambda)}{(\nu+2\lambda)^{3/2}} \qquad \beta_2 = 3+\frac{12(\nu+4\lambda)}{(\nu+2\lambda)^2}$$

$$M(t) = (1-2t)^{-\nu/2}\exp\left[\frac{\lambda t}{1-2t}\right] \qquad \phi(t) = (1-2it)^{-\nu/2}\exp\left[\frac{\lambda it}{1-2it}\right]$$

Noncentral F Distribution

$$f(x)=\sum_{i=0}^{\infty}\frac{\Gamma\left(\frac{2i+\nu_1+\nu_2}{2}\right)\left(\frac{\nu_1}{\nu_2}\right)^{(2i+\nu_1)/2}x^{(2i+\nu_1-2)/2}e^{-\lambda/2}\left(\frac{\lambda}{2}\right)}{\Gamma\left(\frac{\nu_2}{2}\right)\Gamma\left(\frac{2i+\nu_1}{2}\right)\nu_1!\left(1+\frac{\nu_1}{\nu_2}x\right)^{(2i+\nu_1+\nu_2)/2}}$$

$$x>0 \qquad \nu_1,\nu_2\in N \qquad \lambda>0$$

$$\mu=\frac{(\nu_1+\lambda)\nu_2}{(\nu_2-2)\nu_1}, \quad \nu_2>2$$

$$\sigma^2=\frac{(\nu_1+\lambda)^2+2(\nu_1+\lambda)\nu_2^2}{(\nu_2-2)(\nu_2-4)v_1^2}-\frac{(\nu_1+\lambda)^2\nu_2^2}{(\nu_2-2)^2\nu_1^2}, \quad \nu_2>4$$

Noncentral t Distribution

$$f(x)=\frac{\nu^{\nu/2}}{\Gamma\left(\frac{\nu}{2}\right)}\frac{e^{-\delta^2/2}}{\sqrt{\pi}(\nu+x^2)^{(\nu+1)/2}}\sum_{i=0}^{\infty}\Gamma\left(\frac{\nu+i+1}{2}\right)\left(\frac{\delta^i}{i!}\right)\left(\frac{2x^2}{\nu+x^2}\right)^{i/2}$$

$$-\infty<x<\infty; \quad -\infty<\delta<\infty; \quad \nu\in N$$

$$\mu_r'=c_r\frac{\Gamma\left(\frac{\nu-r}{2}\right)\nu^{r/2}}{2^{r/2}\Gamma\left(\frac{\nu}{2}\right)}, \quad \nu>r, \qquad c_{2r-1}=\sum_{i=1}^{r}\frac{(2r-1)!\delta^{2r-1}}{(2i-1)!(r-i)!2^{r-i}},$$

$$c_{2r}=\sum_{i=0}^{r}\frac{(2r)!\delta^{2i}}{(2i)!(r-i)!2^{r-i}}, \qquad r=1,2,3,\ldots$$

Normal Distribution

$$f(x)=\frac{1}{\sigma\sqrt{2\pi}}\exp\left[-\frac{(x-\mu)^2}{2\sigma^2}\right]$$

$$-\infty<x<\infty; \quad -\infty<\mu<\infty; \quad \sigma>0$$

$$\mu=\mu \qquad \sigma^2=\sigma^2 \qquad \beta_1=0 \qquad \beta_2=3 \qquad M(t)=\exp\left[\mu t+\frac{t^2\sigma^2}{2}\right]$$

$$\phi(t)=\exp\left[\mu i t-\frac{t^2\sigma^2}{2}\right]$$

Pareto Distribution

$$f(x) = \theta a^{\theta}/x^{\theta+1} \qquad x \ge a \qquad \theta > 0 \qquad a > 0$$

$$\mu = \frac{\theta a}{\theta - 1}, \quad \theta > 1 \qquad \sigma^2 = \frac{\theta a^2}{(\theta-1)^2(\theta-2)}, \qquad \theta > 2$$

$M(t)$ does not exist.

Rayleigh Distribution

$$f(x) = \frac{x}{\sigma^2}\exp\left[-\frac{x^2}{2\sigma^2}\right] \qquad x \ge 0 \qquad \sigma = 0$$

$$\mu = \sigma\sqrt{\pi/2} \qquad \sigma^2 = 2\sigma^2\left(1 - \frac{\pi}{4}\right) \qquad \beta_1 = \frac{\sqrt{\pi}}{4}\frac{(\pi-3)}{\left(1-\frac{\pi}{4}\right)^{3/2}}$$

$$\beta_2 = \frac{2 - \frac{3}{16}\pi^2}{\left(1 - \frac{\pi}{4}\right)^2}$$

t Distribution

$$f(x) = \frac{1}{\sqrt{\pi\nu}}\frac{\Gamma\left(\frac{\nu+1}{2}\right)}{\Gamma\frac{\nu}{2}}\left(1 + \frac{x^2}{\nu}\right)^{-(\nu+1)/2} \qquad -\infty < x < \infty \qquad \nu \in N$$

$$\mu = 0, \quad \nu \ge 2 \qquad \sigma^2 = \frac{\nu}{\nu - 2}, \quad \nu \ge 3 \qquad \beta_1 = 0, \quad \nu \ge 4$$

$$\beta_2 = 3 + \frac{6}{\nu - 4}, \quad \nu \ge 5$$

$M(t)$ does not exist. $\phi(t) = \dfrac{\sqrt{\pi}\Gamma\left(\frac{\nu}{2}\right)}{\Gamma\left(\frac{\nu+1}{2}\right)} \displaystyle\int_{-\infty}^{\infty} \frac{e^{itz\sqrt{\nu}}}{\left(1+z^2\right)^{(\nu+1)/2}}dz$

Triangular Distribution

$$f(x) = \begin{cases} 0 & x \le a \\ 4(x-a)/(b-a)^2 & a < x \le (a+b)/2 \\ 4(b-x)/(b-a)^2 & (a+b)/2 < x < b \\ 0 & x \ge b \end{cases}$$

$$-\infty < a < b < \infty$$

$$\mu = \frac{a+b}{2} \qquad \sigma^2 = \frac{(b-a)^2}{24} \qquad \beta_1 \doteq 0 \qquad \beta_2 = \frac{12}{5}$$

$$M(t) = -\frac{4\left(e^{at/2} - e^{bt/2}\right)^2}{t^2(b-a)^2} \qquad \phi(t) = \frac{4\left(e^{ait/2} - e^{bit/2}\right)^2}{t^2(b-a)^2}$$

Uniform Distribution

$$f(x) = \frac{1}{b-a} \qquad a \le x \le b \qquad -\infty < a < b < \infty$$

$$\mu = \frac{a+b}{2} \qquad \sigma^2 = \frac{(b-a)^2}{12} \qquad \beta_1 = 0 \qquad \beta_2 = \frac{9}{5}$$

$$M(t) = \frac{e^{bt} - e^{at}}{(b-a)t} \qquad \phi(t) = \frac{e^{bit} - e^{ait}}{(b-a)it}$$

Weibull Distribution

$$f(x) = \frac{\alpha}{\beta^\alpha} x^{\alpha-1} e^{-(x/\beta)^\alpha} \qquad x \ge 0 \qquad \alpha, \beta > 0$$

$$\mu = \beta\Gamma\left(1+\frac{1}{\alpha}\right) \qquad \sigma^2 = \beta^2\left[\Gamma\left(1+\frac{2}{\alpha}\right) - \Gamma^2\left(1+\frac{1}{\alpha}\right)\right]$$

$$\beta_1 = \frac{\Gamma\left(1+\frac{3}{\alpha}\right) - 3\Gamma\left(1+\frac{1}{\alpha}\right)\Gamma\left(1+\frac{2}{\alpha}\right) + 2\Gamma^3\left(1+\frac{1}{\alpha}\right)}{\left[\Gamma\left(1+\frac{2}{\alpha}\right) - \Gamma^2\left(1+\frac{1}{\alpha}\right)\right]^{3/2}}$$

$$\beta_2 = \frac{\Gamma\left(1+\frac{4}{\alpha}\right) - 4\Gamma\left(1+\frac{1}{\alpha}\right)\Gamma\left(1+\frac{3}{\alpha}\right) + 6\Gamma^2\left(1+\frac{1}{\alpha}\right)\Gamma\left(1+\frac{2}{\alpha}\right) - 3\Gamma^4\left(1+\frac{1}{\alpha}\right)}{\left[\Gamma\left(1+\frac{2}{\alpha}\right) - \Gamma^2\left(1+\frac{1}{\alpha}\right)\right]^2}$$

Distribution Parameters

Average

$$\bar{x} = \frac{1}{n}\sum_{i=1}^{n} x_i$$

Variance

$$s^2 = \frac{1}{n-1}\sum_{i=1}^{n}(x_i - \bar{x})^2$$

Standard Deviation

$$s = \sqrt{s^2}$$

Standard Error

$$\frac{s}{\sqrt{n}}$$

Skewness

(Missing if s = 0 or $n < 3$.)

$$\frac{n\sum_{i=1}^{n}(x_i - \bar{x})^3}{(n-1)(n-2)s^3}$$

Standardized Skewness

$$\frac{\text{skewness}}{\sqrt{\frac{6}{n}}}$$

Kurtosis

(Missing if $s = 0$ or $n < 4$.)

$$\frac{n(n+1)\sum_{i=1}^{n}(x_i - \bar{x})^4}{(n-1)(n-2)(n-3)s^4} - \frac{3(n-1)^2}{(n-2)(n-3)}$$

Standardized Kurtosis

$$\frac{\text{Kurtosis}}{\sqrt{\frac{24}{n}}}$$

Weighted Average

$$\frac{\sum_{i=1}^{n} x_i w_i}{\sum_{i=1}^{n} w_i}$$

Estimation and Testing

100(1-α)% Confidence Interval for Mean

$$\bar{x} \pm t_{n-1;\alpha/2} \frac{s}{\sqrt{n}}$$

100(1-α)% Confidence Interval for Variance

$$\left[\frac{(n-1)s^2}{\chi^2_{n-1;\alpha/2}}, \frac{(n-1)s^2}{\chi^2_{n-1;1-\alpha/2}} \right]$$

100(1-α)% Confidence Interval for Difference in Means

Equal Variance

$$(\bar{x}_1 - \bar{x}_2) \pm t_{n_1+n_2-2;\alpha/2}\, s_p \sqrt{\frac{1}{n_1} + \frac{1}{n_2}}$$

where

$$s_p = \sqrt{\frac{(n_1-1)s_1^2 + (n_2-1)s_2^2}{n_1+n_2-2}}$$

Unequal Variance

$$\left[(\bar{x}_1 - \bar{x}_2) \pm t_{m;\alpha/2} \sqrt{\frac{s_1^2}{n_1} + \frac{s_2^2}{n_2}} \right]$$

where

$$\frac{1}{m} = \frac{c^2}{n_1-1} + \frac{(1-c)^2}{n_2-1}$$

and

$$c = \frac{\frac{s_1^2}{n_1}}{\frac{s_1^2}{n_1} + \frac{s_2^2}{n_2}}$$

100(1-α)% Confidence Interval for Ratio of Variances

$$\left(\frac{s_1^2}{s_2^2}\right)\left(\frac{1}{F_{n_1-1,n_2-1;\alpha/2}}\right), \left(\frac{s_1^2}{s_2^2}\right)\left(\frac{1}{F_{n_1-1,n_2-1;\alpha/2}}\right)$$

Normal Probability Plot

The data is sorted from the smallest to the largest value to compute order statistics. A scatter plot is then generated where

$$\text{Horizontal position} = x_{(i)}$$

$$\text{Vertical position} = \Phi\left(\frac{i-3/8}{n+1/4}\right)$$

The labels for the vertical axis are based upon the probability scale using

$$100\left(\frac{i-3/8}{n+1/4}\right)$$

Comparison of Poisson Rates

$$n_j = \#\text{of events in sample } j$$

$$t_j = \text{length of sample } j$$

$$\text{Rate estimates: } r_j = \frac{n_j}{t_j}$$

$$\text{Rate ratio: } \frac{r_1}{r_2}$$

$$\text{Test statistic } z = \max\left(0, \frac{\left|n_1 - \frac{(n_1+n_2)}{2}\right| - \frac{1}{2}}{\sqrt{\frac{(n_1+n_2)}{4}}}\right)$$

where z follows the standard normal distribution.

Distribution Functions-Parameter Estimation

Bernoulli

$$\hat{p} = \bar{x}$$

Binomial

$$\hat{p} = \frac{\bar{x}}{n}$$

where n is the number of trials

Discrete Uniform

$$\hat{a} = \min x_i$$

$$\hat{b} = \max x_i$$

Geometric

$$\hat{p} = \frac{1}{1+\bar{x}}$$

Negative Binomial

$$\hat{p} = \frac{k}{\bar{x}}$$

where k = the number of successes

Poisson

$$\hat{\beta} = \bar{x}$$

Beta

$$\hat{\alpha} = \bar{x}\left[\frac{\bar{x}(1-\bar{x})}{s^2} - 1\right]$$

$$\hat{\beta} = (1-\bar{x})\left(\frac{\bar{x}(1-\bar{x})}{s^2} - 1\right)$$

Chi-Square

$$\text{d.f. } \bar{\nu} = \bar{x}$$

Erlang

$$\hat{\alpha} = \text{round } (\hat{\alpha} \text{ from gamma})$$

$$\hat{\beta} = \frac{\hat{\alpha}}{x}$$

Exponential

$$\hat{\beta} = \frac{1}{\bar{x}}$$

Note
System displays $1/\hat{\beta}$.

F

$$\text{num d.f.: } \hat{\nu} = \frac{2\hat{w}^3 - 4\hat{w}^2}{\left(s^2(\hat{w}-2)^2(\hat{w}-4)\right) - 2\hat{w}^2}$$

$$\text{den. d.f.: } \hat{w} = \frac{\max(1, 2\bar{x})}{-1+\bar{x}}$$

Gamma

$$R = \log\left(\frac{\text{arithmetic mean}}{\text{geometric mean}}\right)$$

If $0 < R \leq 0.5772$,

$$\hat{\alpha} = R^{-1}\left(0.5000876 + 0.1648852\, R - 0.0544274\, R\right)^2$$

or if $R > 0.5772$,

$$\hat{\alpha} = R^{-1}\left(17.79728 + 11.968477\,R + R^2\right)^{-1}\left(8.898919 + 9.059950\,R + 0.9775373\,R^2\right)$$

$$\hat{\beta} = \hat{\alpha}/\bar{x}$$

This is an approximation of the Method of Maximum Likelihood solution from Johnson and Kotz (1970).

Log Normal

$$\hat{\mu} = \frac{1}{n}\sum_{i=1}^{n} \log x_i$$

$$\hat{\alpha} = \sqrt{\frac{1}{n-1}\sum_{i=1}^{n}\left(\log x_i - \hat{\mu}\right)^2}$$

$$\text{means } \exp\left(\hat{\mu} + \hat{\alpha}^2/2\right)$$

$$\text{Standard deviation } \sqrt{\exp\left(2\hat{\mu} + \hat{\alpha}^2\right)\left[\exp\left(\hat{\alpha}^2\right) - 1\right]}$$

Normal

$$\hat{\mu} = \bar{x}$$

$$\hat{\sigma} = s$$

Student's t

If $s^2 \leq 1$ or if $\hat{v} \leq 2$, then the system indicates that the data are inappropriate.

$$s^2 = \frac{\sum_{i=1}^{n} x_i^2}{n}$$

$$\hat{v} = \frac{2s^2}{-1 + s^2}$$

Triangular

$$\hat{a} = \min x_i$$

$$\hat{c} = \max x_i$$

$$\hat{b} = 3\overline{x} - \hat{a} - \overline{x}$$

Uniform

$$\hat{a} = \min x_i$$

$$\hat{b} = \max x_i$$

Weibull

Solves the simultaneous equations

$$\hat{\alpha} = \frac{n}{\left[\left(\frac{1}{\hat{\beta}}\right)\sum_{i=1}^{n} x_i^{\hat{a}} \log x_i - \sum_{i=1}^{n} \log x_i\right]}$$

$$\hat{\beta} = \left(\frac{\sum_{i=1}^{n} x_i^{\hat{\alpha}}}{n}\right)^{\frac{1}{\hat{\alpha}}}$$

Chi-Square Test for Distribution Fitting

Divide the range of data into nonoverlapping classes. The classes are aggregated at each end to ensure that classes have an expected frequency of at least 5.

O_i = observed frequency in class i

E_i = expected frequency in class i from fitted distribution

k = number of classes after aggregation

Test statistic

$$\chi^2 = \sum_{i=1}^{k} \frac{(O_i - E_i)^2}{E_i}$$

follows a chi-square distribution with the degrees of freedom equal to (k–1) – (Number of estimated parameters).

Kolmogorov–Smirnov Test

$$D_n^+ = \max\left\{\frac{i}{n} - \hat{F}(x_i)\right\}$$

$$1 \leq i \leq n$$

$$D_n^- = \max\left\{\hat{F}(x_i) - \frac{i-1}{n}\right\}$$

$$1 \leq i \leq n$$

$$D_n = \max\left\{D_n^+, D_n^-\right\}$$

where $\hat{F}(x_i) =$ estimated cumulative distribution at x_i.

ANOVA

Notations

k = number of treatments

n_t = number of observations for treatment t

$$\bar{n} = n/k = \text{average treatment size}$$

$$n = \sum_{t=1}^{k} n_t$$

$$x_{it} = i^{th} \text{ observation in treatment } t$$

$$\bar{x}_t = \text{treatment mean} = \frac{\sum_{i=1}^{n_t} x_{it}}{n_t}$$

$$s_t^2 = \text{treatment variance} = \frac{\sum_{i=1}^{n_t} \left(x_{it} - \bar{x}_t\right)^2}{n_t - 1}$$

$$\text{MSE} = \text{mean square error} = \frac{\sum_{t=1}^{k}(n_t - 1)s_t^2}{\left(\sum_{t=1}^{k} n_t\right) - k}$$

$$\text{df} = \text{degrees of freedom for the error term} = \left(\sum_{t=1}^{k} n_t\right) - k$$

Standard Error (Internal)

$$\sqrt{\frac{s_t^2}{n_t}}$$

Standard Error (Pooled)

$$\sqrt{\frac{MSE}{n_t}}$$

Interval Estimates

$$\bar{x}_t \pm M\sqrt{\frac{MSE}{n_t}}$$

where confidence interval is

$$M = t_{n-k;\alpha/2}$$

and LSD interval is

$$M = \frac{1}{\sqrt{2}} t_{n-k;\,\alpha/2}$$

Tukey Interval

$$M = \frac{1}{2} q_{n-k,\,k;\,\alpha}$$

where $q_{n-k,k;\alpha} =$ the value of the studentized range distribution with $n - k$ degrees of freedom and k samples such that the cumulative probability equals $1 - \alpha$.

Scheffe Interval

$$M = \frac{\sqrt{k-1}}{\sqrt{2}} \sqrt{F_{k-1,\, n-k;\, \alpha}}$$

Cochran C-Test

Follow F distribution with $\bar{n}-1$ and $(\bar{n}-1)(k-1)$ degrees of freedom.
Test statistic:

$$F = \frac{(k-1)C}{1-C}$$

where

$$C = \frac{\max s_t^2}{\sum_{t=1}^{k} s_t^2}$$

Bartlett Test

Test statistic:

$$B = 10^{\frac{M}{(n-k)}}$$

$$M = (n-k)\log_{10} MSE - \sum_{t=1}^{k} (n_t - 1)\log_{10} s_t^2$$

The significance test is based on

$$\frac{M(1n\,10)}{1 + \frac{1}{3(k-1)}\left[\sum_{t=1}^{k} \frac{1}{(n_t-1)} - \frac{1}{N-k}\right]^{X_{k-1}^2}}$$

which follows a chi-square distribution with k – 1 degrees of freedom.

Hartley's Test

$$H = \frac{\max\left(s_t^2\right)}{\min\left(s_t^2\right)}$$

Kruskal-Wallis Test

Average rank of treatment:

$$\bar{R}_t = \frac{\sum_{i=1}^{n_t} R_{it}}{n_t}$$

If there are no ties, then the test statistic is

$$w = \left(\frac{12}{n} \sum_{i=1}^{k} n_t \bar{R}_t^{\,2} \right) - 3(n+1)$$

Adjustment for Ties

Let u_j be the number of observations tied at any rank for $j = 1,2,3, \ldots, m$, where m is the number of unique values in the sample.

$$W = \frac{w}{1 - \dfrac{\sum_{j=1}^{m} u_j^3 - \sum_{j=1}^{m} u_j}{n(n^2-1)}}$$

Significance level: W follows a chi-square distribution with $k - 1$ degrees of freedom.

Freidman Test

$$X_{it} = \text{observation in the } i^{\text{th}} \text{row, } t^{th} \text{ column}$$

$$i = 1, 2, \ldots, n; \quad t = 1, 2, \ldots, k$$

$$R_{it} = \text{rank of } X_{it} \text{ within its row}$$

where n is the common treatment size (all treatment sizes must be the same for this test).

$$R_t = \sum_{i=1}^{n} R_{i\,t}$$

$$\text{average rank } \bar{R}_t = \frac{\sum_{i=1}^{n_t} R_{i\,t}}{n_t}$$

where data are ranked within each row separately.

$$\text{Test statistic } Q = \frac{12S(k-1)}{nk(k^2-1)-\left(\sum u^3 - \sum u\right)}$$

where

$$S = \left(\sum_{t=1}^{k} R_i^2\right) - \frac{n^2k(k+1)^2}{4}$$

Q follows a chi-square distribution with k degrees of freedom.

Regression

Notation

$\underset{\sim}{Y} =$ vector of n observation for the dependent variable ~
$\underset{\sim}{X} = n$ by p matrix of observations for independent variables, including constant term, if any ~ indicates a variable is a vector or matrix

$$\bar{Y} = \frac{\sum_{i=1}^{n} Y_i}{n}$$

Regression Statistics

Estimated Coefficients

Note: Estimated by a modified Gram-Schmidt orthogonal decomposition with a tolerance equal to 1.0E−08.

$$\underset{\sim}{b} = (\underset{\sim}{X}'\underset{\sim}{X})^{-1} \underset{\sim}{X}\underset{\sim}{Y}$$

Standard Errors

$$S(\underset{\sim}{b}) = \sqrt{\text{diagonal elements of } (\underset{\sim}{X}'\underset{\sim}{X})^{-1}\,\text{MSE}}$$

where

$$SSE = \underset{\sim}{Y}'\underset{\sim}{Y} - \underset{\sim}{b}'\underset{\sim}{X}'\underset{\sim}{Y}$$

$$MSE = \frac{SSE}{n-p}$$

t-Values

$$\underset{\sim}{t} = \frac{\underset{\sim}{b}}{S(\underset{\sim}{b})}$$

Significance Level

t-Values follow the Student's t distribution with $n - p$ degrees of freedom.

R-Squared

$$R^2 = \frac{SSTO - SSE}{SSTO}$$

where

$$SSTO = \begin{cases} \underset{\sim}{Y}' - n\bar{Y}^2 & \text{if constant} \\ \underset{\sim}{Y}\underset{\sim}{Y} & \text{if no constant} \end{cases}$$

Note: When the no constant option is selected, the total sum of square is uncorrected for the mean. Thus, the R^2 value is of little use, since the sum of the residuals is not zero.

Adjusted R-Squared

$$1 - \left(\frac{n-1}{n-p}\right)\left(1 - R^2\right)$$

Standard Error of Estimate

$$SE = \sqrt{MSE}$$

Predicted Values

$$\hat{\underset{\sim}{Y}} = \underset{\sim}{X}\underset{\sim}{b}$$

Residuals

$$\underset{\sim}{e} = \underset{\sim}{Y} - \hat{\underset{\sim}{Y}}$$

Durbin–Watson Statistic

$$D = \frac{\sum_{i=1}^{n-1}(e_{i+1} - e_i)^2}{\sum_{i=1}^{n} e_i^2}$$

Mean Absolute Error

$$\frac{\left(\sum_{i=1}^{n} |e_i|\right)}{n}$$

Predictions

$\underset{\sim}{X}_h = m$ by p matrix of independent variables for m predictions.

Predicted Value

$$\hat{\underset{\sim}{Y}}_h = \underset{\sim}{X}_h \underset{\sim}{b}$$

Standard Error of Predictions

$$S\left(\hat{\underset{\sim}{Y}}_{h(new)}\right) = \sqrt{\text{diagonal elements of MSE } \left(1 + \underset{\sim}{X}_h \left(\underset{\sim}{X}'\underset{\sim}{X}\right)^{-1} \underset{\sim}{X}'_h\right)}$$

Standard Error of Mean Response

$$S\left(\hat{\underset{\sim}{Y}}_h\right) = \sqrt{\text{diagonal elements of MSE } \left(\underset{\sim}{X}_h \left(\underset{\sim}{X}'\underset{\sim}{X}\right)^{-1} \hat{\underset{\sim}{X}}_h\right)}$$

Prediction Matrix Results

Column 1 = index numbers of forecasts

$$2 = \hat{\underset{\sim}{Y}}_h$$

$$3 = S\left(\hat{\underset{\sim}{Y}}_{h(new)}\right)$$

$$4 = \left(\hat{\underset{\sim}{Y}}_h - t_{n-p,\,\alpha/2}\, S\left(\hat{\underset{\sim}{Y}}_{h(new)}\right)\right)$$

$$5 = \left(\hat{\underset{\sim}{Y}}_h + t_{n-p,\alpha/2}\, S\left(\hat{\underset{\sim}{Y}}_{h(new)}\right)\right)$$

$$6 = \hat{\underset{\sim}{Y}}_h - t_{n-p,\alpha/2}\, S\left(\hat{\underset{\sim}{Y}}_h\right)$$

$$7 = \hat{\underset{\sim}{Y}}_h + t_{n-p,\alpha/2}\, S\left(\hat{\underset{\sim}{Y}}_h\right)$$

Nonlinear Regression

$F(X,\hat{\beta})$ are values of nonlinear function using parameter estimates $\hat{\beta}$.

Estimated Coefficients

Obtained by minimizing the residual sum of squares using a search procedure suggested by Marquardt. This is a compromise between Gauss-Newton and steepest descent methods. The user specifies

- Initial estimates $\underset{\sim}{\beta}_0$
- Initial value of Marquardt parameter λ, which is modified at each iteration. As $\lambda \to 0$, the procedure approaches Gauss-Newton $\lambda \to \infty$, and procedure approaches steepest descent.
- Scaling factor used to multiply Marquardt parameter after each iteration
- Maximum value of Marquardt parameter

Partial derivatives of F with respect to each parameter are estimated numerically.

Standard Errors

Standard errors are estimated from the residual sum of squares and partial derivatives.

Ratio

$$\text{ratio} = \frac{\text{coefficient}}{\text{standard error}}$$

R-Squared

$$R^2 = \frac{\text{SSTO-SSE}}{\text{SSTO}} \quad \text{where}$$

$$\text{SSTO} = \underset{\sim}{Y}'\underset{\sim}{Y} - n\bar{Y}^2$$

$$\text{SSE} = \text{residual sum of squares}$$

Ridge Regression

Additional Notation

$\underset{\sim}{Z}$ = matrix of independent variables standardized so that $\underset{\sim}{Z}'\underset{\sim}{Z}$ equals the correlation matrix
θ = value of the ridge parameter

Parameter Estimates

$$\underset{\sim}{b}(\theta) = \left(\underset{\sim}{Z}'\underset{\sim}{Z} + \theta \mathrm{I}_p\right)^{-1} \underset{\sim}{Z}'\underset{\sim}{Y}$$

where I_p is a $p \times p$ identity matrix.

Quality Control

For all quality control formulas

k = number of subgroups
n_j = number of observations in subgroup j, $j = 1, 2, \ldots, k$
x_{ij} = ith observation in subgroup j

All following formulas for quality control assume 3-sigma limits. If other limits are specified, the formulas are adjusted proportionally based on sigma for the selected limits. Also, average sample size is used unless otherwise specified.

Subgroup Statistics

Subgroup Means

$$\bar{x}_j = \frac{\sum_{i=1}^{n_j} x_{ij}}{n_j}$$

Subgroup Standard Deviations

$$s_j = \sqrt{\frac{\sum_{i=1}^{n_j} \left(x_{ij} - \bar{x}_j\right)^2}{\left(n_j - 1\right)}}$$

Subgroup Range

$$R_j = \max\left\{x_{ij} \middle| 1 \le i \le n_j\right\} - \min\left\{x_{ij} \middle| 1 \le i \le n_j\right\}$$

X Bar Charts

Compute

$$\bar{\bar{x}} = \frac{\sum_{j=1}^{k} n_i \bar{x}_j}{\sum_{j=1}^{k} n_i}$$

$$\bar{R} = \frac{\left(\sum_{j=1}^{k} n_i R_j\right)}{\sum_{j=1}^{k} n_i}$$

$$s_p = \sqrt{\frac{\sum_{j=1}^{k} (n_j - 1)s_j^2}{\sum_{j=1}^{k} (n_j - 1)}}$$

$$\bar{n} = \frac{1}{k}\sum_{j=1}^{k} n_i$$

For a chart based on range:

$$UCL = \bar{\bar{x}} + A_2\bar{R}$$

$$LCL = \bar{\bar{x}} - A_2\bar{R}$$

For a chart based on sigma:

$$UCL = \bar{\bar{x}} + \frac{3s_p}{\sqrt{\bar{n}}}$$

$$LCL = \bar{\bar{x}} - \frac{3s_p}{\sqrt{\bar{n}}}$$

For a chart based on known sigma:

$$UCL = \bar{\bar{x}} + 3\frac{\sigma}{\sqrt{\bar{n}}}$$

$$LCL = \bar{\bar{x}} - 3\frac{\sigma}{\sqrt{\bar{n}}}$$

If other than 3-sigma limits are used, such as 2-sigma limits, all bounds are adjusted proportionately. If average sample size is not used, then uneven bounds are displays based on $1/\sqrt{n_j}$ rather than $1/\sqrt{\bar{n}}$.

If the data is normalized, each observation is transformed according to

$$z_{ij} = \frac{x_{ij} - \bar{\bar{x}}}{\hat{\alpha}}$$

where $\hat{\alpha}$ = estimated standard deviation.

Capability Ratios

Note: The following indices are useful only when the control limits are placed at the specification limits. To override the normal calculations, specify a subgroup size of one and select the "known standard deviation" option. Then enter the standard deviation as half of the distance between the USL and LSL. Change the position of the center line to be the midpoint of the USL and LSL, and specify the upper and lower control line at one sigma.

$$C_P = \frac{USL - LSL}{6\hat{\alpha}}$$

$$C_R = \frac{1}{C_P}$$

$$C_{PK} = \min\left(\frac{USL - \bar{\bar{x}}}{3\hat{\alpha}}, \frac{\bar{\bar{x}} - LSL}{3\hat{\alpha}}\right)$$

R Charts

$$CL = \bar{R}$$

$$UCL = D_4\bar{R}$$

$$LCL = Max\left(0, D_3\bar{R}\right)$$

S Charts

$$CL = s_P$$

$$UCL = s_P\sqrt{\frac{\chi^2_{\bar{n}-1;\alpha}}{\bar{n}-1}}$$

$$LCL = s_P\sqrt{\frac{\chi^2_{\bar{n}-1;\alpha}}{\bar{n}-1}}$$

C Charts

$$\bar{c} = \sum u_j \qquad UCL = \bar{c} + 3\sqrt{\bar{c}}$$

$$\sum n_j \qquad LCL = \bar{c} - 3\sqrt{\bar{c}}$$

where u_j is the number of defects in the jth sample

U Charts

$$\bar{u} = \frac{\text{number of defects in all samples}}{\text{number of units in all samples}} = \frac{\sum u_j}{\sum n_j}$$

$$UCL = \bar{u} + \frac{3\sqrt{\bar{u}}}{\sqrt{\bar{n}}}$$

$$LCL = \bar{u} - \frac{3\sqrt{\bar{u}}}{\sqrt{\bar{n}}}$$

P Charts

$$p = \frac{\text{number of defective units}}{\text{number of units inspected}}$$

$$\bar{p} = \frac{\text{number of defectives in all samples}}{\text{number of units in all samples}} = \frac{\sum p_j n_j}{\sum n_j}$$

$$UCL = \bar{p} + \frac{3\sqrt{\bar{p}(1-\bar{p})}}{\sqrt{\bar{n}}}$$

$$LCL = \bar{p} - \frac{3\sqrt{\bar{p}(1-\bar{p})}}{\sqrt{\bar{n}}}$$

NP Charts

$$\bar{p} = \frac{\sum d_j}{\sum n_j},$$

where d_j is the number of defectives in the jth sample.

$$UCL = \bar{n}\,\bar{p} + 3\sqrt{\bar{n}\,\bar{p}(1-\bar{p})}$$

$$LCL = \bar{n}\,\bar{p} - 3\sqrt{\bar{n}\,\bar{p}(1-\bar{p})}$$

CuSum Chart for the Mean

Control mean = μ
Standard deviation = α
Difference to detect = Δ

Plot cumulative sums C_t versus t where

$$C_t = \sum_{i=1}^{t} (\bar{x}_i - \mu) \quad \text{for } t = 1, 2, \ldots, n$$

The V-mask is located at distance

$$d = \frac{2}{\Delta}\left[\frac{\alpha^2/\bar{n}}{\Delta}\, 1n\, \frac{1-\beta}{\alpha/2}\right]$$

in front of the last data point.

Angle of mast $= 2\tan^{-1}\frac{\Delta}{2}$

Slope of the lines $= \pm\frac{\Delta}{2}$

Multivariate Control Charts

$\underset{\sim}{X}$ = matrix of n rows and k columns containing n observations for each of k variable

S = sample covariance matrix

$\underset{\sim}{X}_t$ = observation vector at time t

$\underset{\sim}{\bar{X}}$ = vector of column average

Then,

$$T_t^2 = \left(\underset{\sim}{X}_t - \underset{\sim}{\bar{X}}\right) S^{-1} \left(\underset{\sim}{X}_t - \underset{\sim}{\bar{X}}\right)$$

$$UCL = \left(\frac{k(n-1)}{n-k}\right) F_{k,n-k;\alpha}$$

Time Series Analysis

Notation

$$x_t \text{ or } y_t = \text{observation at time } t, t = 1, 2, \ldots, n$$

n = number of observations

Autocorrelation at Lag k

$$r_k = \frac{c_k}{c_0}$$

where

$$c_k = \frac{1}{n}\sum_{t=1}^{n-k}(y_t - \bar{y})(y_{t+k} - \bar{y})$$

and

$$\bar{y} = \frac{\left(\sum_{t=1}^{n} y_t\right)}{n}$$

$$\text{standard error} = \sqrt{\frac{1}{n}\left\{1 + 2\sum_{v=1}^{k-1} r_v^2\right\}}$$

Partial Autocorrelation at Lag k

$\hat{\theta}_{kk}$ is obtained by solving the Yule-Walker equations:

$$r_j = \hat{\theta}_{k1} r_{j-1} + \hat{\theta}_{k2} r_{j-2} + \cdots + \hat{\theta}_{k(k-1)} r_{j-k+1} + \hat{\theta}_{kk} r_{j-k}$$

$j = 1, 2, \ldots, k$

$$\text{Standard error} = \sqrt{\frac{1}{n}}$$

Cross Correlation at Lag k

x = input time series

y = output time series

$$r_{xy}(k) = \frac{c_{xy}(k)}{s_x s_y} \quad k = 0, \pm 1, \pm 2, \ldots$$

where

$$c_{xy}(k) = \begin{cases} \dfrac{1}{n} \sum_{t=1}^{n-k} (x_t - \bar{x})(y_{t+k} - \bar{y}) & k = 0, 1, 2, \ldots \\ \dfrac{1}{n} \sum_{t=1}^{n+k} (x_t - \bar{x})(y_{t-k} - \bar{y}) & k = 0, -1, -2, \ldots \end{cases}$$

and

$$S_x = \sqrt{c_{xx}(0)}$$

$$S_y = \sqrt{c_{yy}(0)}$$

Box-Cox

$$yt = \frac{(y + \lambda_2)^{\lambda_1} - 1}{\lambda_1 g^{(\lambda_1 - 1)}} \qquad \text{if } \lambda_1 > 0$$

$$yt = g\, 1n(y + \lambda_2) \qquad \text{if } \lambda_1 = 0$$

where g is the sample geometric mean $(y + \lambda_2)$.

Periodogram (Computed Using Fast Fourier Transform)

If n is odd

$$I(f_1) = \frac{n}{2}\left(a_i^2 + b_i^2\right) \qquad i = 1, 2, \ldots, \left[\frac{n-1}{2}\right]$$

where

$$a_i = \frac{2}{n}\sum_{t=1}^{n} t_t \cos 2\pi f_i t$$

$$b_i = \frac{2}{n}\sum_{t=1}^{n} y_t \sin 2\pi f_i t$$

$$f_i = \frac{i}{n}$$

If n is even, an additional term is added

$$I(0.5) = n\left(\frac{1}{n}\sum_{t=1}^{n}(-1)^t Y_t\right)^2$$

Categorical Analysis

Notation

r = number of rows in table
c = number of columns in table
f_{ij} = frequency in position (row i, column j)
xi = distinct values of row variable arranged in ascending order; $i = 1, \ldots, r$
y_j = distinct values of column variable arranged in ascending order, $j = 1, \ldots, c$

Totals

$$R_j = \sum_{j=1}^{c} f_{ij} \qquad C_j = \sum_{i=1}^{r} f_{ij}$$

$$N = \sum_{i=1}^{r}\sum_{j=1}^{c} f_{ij}$$

Note: Any row or column that totals zero is eliminated from the table before calculations are performed.

Chi-Square

$$\chi^2 = \sum_{i=1}^{r}\sum_{j=1}^{c}\frac{\left(f_{ij} - E_{ij}\right)^2}{E_{ij}}$$

where

$$E_{ij} = \frac{R_i C_j}{N} \sim \chi^2_{(r-1)(c-1)}$$

A warning is issued if any $E_{ij} < 2$ or if 20% or more of all $E_{ij} < 5$. For 2 × 2 tables, a second statistic is printed using Yate's continuity correction.

Fisher's Exact Test

Run for a 2 × 2 table, when N is less than or equal to 100. For calculation details, see standard references such as *The Analysis of Contingency Tables* by B. S. Everitt.

Lambda

$$\lambda = \frac{\left(\sum_{j=1}^{c} f_{\max,j} - R_{\max}\right)}{N - R_{\max}} \text{ with rows dependent}$$

$$\lambda = \frac{\left(\sum_{i=1}^{r} f_{i,\max} - C_{\max}\right)}{N - C_{\max}} \text{ with columns dependent}$$

$$\lambda = \frac{\left(\sum_{i=1}^{r} f_{i,\max} + \sum_{j=1}^{c} f_{\max,j} - C_{\max} - R_{\max}\right)}{(2N - R_{\max} - C_{\max})} \text{ when symmetric}$$

where

$f_{i\max}$ = largest value in row i

f_{maxj} = largest value in column j

R_{max} = largest row total

C_{max} = largest column total

Uncertainty Coefficient

$$U_R = \frac{U(R)+U(C)-U(RC)}{U(R)} \text{ with rows dependent}$$

$$U_C = \frac{U(R)+U(C)-U(RC)}{U(C)} \text{ with columns dependent}$$

$$U = 2\left(\frac{U(R)+U(C)-U(RC)}{U(R)+U(C)}\right) \text{when symmetric}$$

where

$$U(R) = -\sum_{i=1}^{r} \frac{R_i}{N} \log \frac{R_i}{N}$$

$$U(C) = -\sum_{j=1}^{c} \frac{C_j}{N} \log \frac{C_j}{N}$$

$$U(RC) = -\sum_{i=1}^{r}\sum_{j=1}^{c} \frac{f_{ij}}{N} \log \frac{f_{ij}}{N} \qquad \text{for } f_{ij} > 0$$

Somer's D

$$D_R = \frac{2(P_C - P_D)}{\left(N^2 - \sum_{j=1}^{c} C_j^2\right)} \text{ with rows dependent}$$

$$D_C = \frac{2(P_C - P_D)}{\left(N^2 - \sum_{i=1}^{r} R_i^2\right)} \text{ with columns dependent}$$

$$D = \frac{4(P_C - P_D)}{\left(N^2 - \sum_{i=1}^{r} R_i^2\right) + \left(N^2 - \sum_{j=1}^{c} C_j^2\right)} \text{ when symmetric}$$

where the number of concordant pairs is

$$P_C = \sum_{i=1}^{r}\sum_{j=1}^{c} f_{ij} \sum_{h<i}\sum_{k<j} f_{hk}$$

and the number of discordant pairs is

$$P_D = \sum_{i=1}^{r}\sum_{j=1}^{c} \mathrm{f}_{ij} \sum_{h<i}\sum_{k>j} \mathrm{f}_{hk}$$

Eta

$$E_R = \sqrt{1 - \frac{SS_{RN}}{SS_R}} \text{ with rows dependent}$$

where the total corrected sum of squares for the rows is

$$SS_R = \sum_{i=1}^{r}\sum_{j-1}^{c} x_i^2 \mathrm{f}_{ij} - \frac{\left(\sum_{i=1}^{r}\sum_{j-1}^{c} x_i \mathrm{f}_{ij}\right)^2}{N}$$

and the sum of squares of rows within categories of columns is

$$SS_{RN} = \sum_{j=1}^{c}\left(\sum_{i=1}^{r} x_i^2 \mathrm{f}_{ij} - \frac{\left(\sum_{i=1}^{r} x_i^2 \mathrm{f}_{ij}\right)^2}{C_j}\right)$$

$$E_C = \sqrt{1 - \frac{SS_{CN}}{SS_C}} \text{ with columns dependent}$$

where the total corrected sum of squares for the columns is

$$SS_C = \sum_{i=1}^{r}\sum_{j=1}^{c} y_i^2 \mathrm{f}_{ij} - \frac{\left(\sum_{i=1}^{r}\sum_{j=1}^{c} y_i \mathrm{f}_{ij}\right)^2}{N}$$

and the sum of squares of columns within categories of rows is

$$SS_{CN} = \sum_{i=1}^{r}\left(\sum_{j=1}^{c} y_i^2 \mathrm{f}_{ij} - \frac{\left(\sum_{j=1}^{c} y_j^2 \mathrm{f}_{ij}\right)^2}{R_i}\right)j$$

Contingency Coefficient

$$C = \sqrt{\frac{\chi^2}{(\chi^2 + N)}}$$

Cramer's V

$$V = \sqrt{\frac{\chi^2}{N}} \text{ for } 2 \times 2 \text{ table}$$

$$V = \sqrt{\frac{\chi^2}{N(m-1)}} \text{ for all others}$$

where $m = \min(r,c)$.

Conditional Gamma

$$G = \frac{P_C - P_D}{P_C + P_D}$$

Pearson's *R*

$$R = \frac{\sum_{j=1}^{c}\sum_{i=1}^{r} x_i y_j f_{ij} - \frac{\left(\sum_{j=1}^{c}\sum_{i=1}^{r} x_i f_{ij}\right)\left(\sum_{j=1}^{c}\sum_{i=1}^{r} y_i f_{ij}\right)}{N}}{\sqrt{SS_R SS_C}}$$

If $R = 1$, no significance is printed. Otherwise, the one-sided significance is based on

$$t = R\sqrt{\frac{N-2}{1-R^2}}$$

Kendall's Tau *b*

$$\tau = \frac{2(P_C - P_D)}{\sqrt{\left(N^2 - \sum_{i=1}^{r} R_i^2\right)\left(N^2 - \sum_{j=1}^{c} C_j^2\right)}}$$

Tau *C*

$$\tau_C = \frac{2m(P_C - P_D)}{(m-1)N^2}$$

Probability Terminology

Experiment—An experiment is an activity or occurrence with an observable result.

Outcome—The result of the experiment.

Sample point—An outcome of an experiment.

Event—An event is a set of outcomes (a subset of the sample space) to which a probability is assigned.

Basic Probability Principles

Consider a random sampling process in which all the outcomes solely depend on chance, that is, each outcome is equally likely to happen. If *S* is a uniform sample space and the collection of desired outcomes is *E*, the probability of the desired outcomes is

$$P(E) = \frac{n(E)}{n(S)}$$

where

$n(E)$ = number of favorable outcomes in *E*

$n(S)$ = number of possible outcomes in *S*

Since E is a subset of S, $0 \le n(E) \le n(S)$, the probability of the desired outcome is $0 \le P(E) \le 1$.

Random Variable

A random variable is a rule that assigns a number to each outcome of a chance experiment. For example:

- A coin is tossed six times. The random variable *X* is the number of tails that are noted. *X* can only take the values 1, 2, …, 6, so *X* is a discrete random variable.

- A light bulb is burned until it burns out. The random variable Y is its lifetime in hours. Y can take any positive real value, so Y is a continuous random variable.

Mean Value $\hat{x}$ or Expected Value μ

The mean value or expected value of a random variable indicates its average or central value. It is a useful summary value of the variable's distribution.

1. If random variable X is a discrete mean value,

$$\hat{x} = x_1 p_1 + x_2 p_2 + \cdots + x_n p_n = \sum_{i=1}^{n} x_1 p_1$$

   where p_i is the probability density.

2. If X is a continuous random variable with probability density function $f(x)$, then the expected value of X is

$$\mu = E(X) = \int_{-\infty}^{+\infty} x f(x) dx$$

   where $f(x)$ is the probability density.

Discrete Distribution Formulas

Probability mass function, $p(x)$

Mean, μ

Variance, s^2

Coefficient of skewness, b_1

Coefficient of kurtosis, b_2

Moment-generating function, $M(t)$

Characteristic function, $f(t)$

Probability-generating function, $P(t)$

Bernoulli Distribution

$$p(x) = p^x q^{x-1} \quad x = 0,1 \quad 0 \le p \le 1 \quad q = 1 - p$$

$$\mu = p \qquad \sigma^2 = pq \qquad \beta_1 = \frac{1-2p}{\sqrt{pq}} \qquad \beta_2 = 3 + \frac{1-6pq}{pq}$$

$$M(t) = q + pe^t \qquad \phi(t) = q + pe^{it} \qquad P(t) = q + pt$$

Beta Binomial Distribution

$$p(x) = \frac{1}{n+1} \frac{B(a+x, b+n-x)}{B(x+1, n-x+1)B(a,b)} \qquad x = 0,1,2,\ldots,n \qquad a > 0 \qquad b > 0$$

$$\mu = \frac{na}{a+b} \qquad \sigma^2 = \frac{nab(a+b+n)}{(a+b)^2(a+b+1)}$$

$B(a,b)$ is the beta function.

Beta Pascal Distribution

$$p(x) = \frac{\Gamma(x)\Gamma(\nu)\Gamma(\rho+\nu)\Gamma(\nu+x-(\rho+r))}{\Gamma(r)\Gamma(x-r+1)\Gamma(\rho)\Gamma(\nu-\rho)\Gamma(\nu+x)} \qquad x = r, r+1, \ldots \quad \nu > p > 0$$

$$\mu = r\frac{\nu-1}{\rho-1}, \quad \rho > 1 \qquad \sigma^2 = r(r+\rho-1)\frac{(\nu-1)(\nu-\rho)}{(\rho-1)^2(\rho-2)}, \quad \rho > 2$$

Binomial Distribution

$$p(x) = \binom{n}{x} p^x q^{n-x} \quad x = 0,1,2,\ldots,n \quad 0 \le p \le 1 \quad q = 1-p$$

$$\mu = np \qquad \sigma^2 = npq \qquad \beta_1 = \frac{1-2p}{\sqrt{npq}} \qquad \beta_2 = 3 + \frac{1-6pq}{npq}$$

$$M(t) = (q + pe^t)^n \qquad \phi(t) = (q + pe^{it})^n \qquad P(t) = (q + pt)^n$$

Discrete Weibull Distribution

$$p(x) = (1-p)^{x^\beta} - (1-p)^{(x+1)^\beta} \qquad x = 0,1,\ldots \quad 0 \le p \le 1 \quad \beta > 0$$

Geometric Distribution

$$p(x) = pq^{1-x} \qquad x = 0,1,2,\ldots \quad 0 \le p \le 1 \quad q = 1-p$$

$$\mu = \frac{1}{p} \qquad \sigma^2 = \frac{q}{p^2} \qquad \beta_1 = \frac{2-p}{\sqrt{q}} \qquad \beta_2 = \frac{p^2+6q}{q}$$

$$M(t) = \frac{p}{1-qe^t} \qquad \phi(t) = \frac{p}{1-qe^{it}} \qquad P(t) = \frac{p}{1-qt}$$

Hypergeometric Distribution

$$p(x) = \frac{\binom{M}{x}\binom{N-M}{n-x}}{\binom{N}{n}} \qquad x = 0,1,2,\ldots,n \qquad x \le M \qquad n-x \le N-M$$

$$n, M, N, \in N; \quad 1 \le n \le N; \quad 1 \le M \le N; \quad N = 1, 2, \ldots$$

$$\mu = n\frac{M}{N} \qquad \sigma^2 = \left(\frac{N-n}{N-1}\right) n\frac{M}{N}\left(1-\frac{M}{N}\right) \qquad \beta_1 = \frac{(N-2M)(N-2n)\sqrt{N-1}}{(N-2)\sqrt{nM(N-M)(N-n)}}$$

$$\beta_2 = \frac{N^2(N-1)}{(N-2)(N-3)nM(N-M)(N-n)}$$

$$\left\{N(N+1)-6n(N-n)+3\frac{M}{N^2}(N-M)\left[N^2(n-2)-Nn^2+6n(N-n)\right]\right\}$$

$$M(t) = \frac{(N-M)!(N-n)!}{N!}F(.,e^t)$$

$$\phi(t) = \frac{(N-M)!(N-n)!}{N!}F(.,e^{it}) \quad P(t) = \left(\frac{N-M}{N}\right)^n F(.,t)$$

$F(a, b, g, x)$ is the hypergeometric function. $\alpha = -n$; $\beta = -M$; $\gamma = N - M - n + 1$

Negative Binomial Distribution

$$p(x) = \binom{x+r-1}{r-1} p^r q^x \qquad x = 0,1,2,\ldots \qquad r = 1,2,\ldots \qquad 0 \le p \le 1 \qquad q = 1-p$$

$$\mu = \frac{rq}{p} \qquad \sigma^2 = \frac{rq}{p^2} \qquad \beta_1 = \frac{2-p}{\sqrt{rq}} \qquad \beta_2 = 3 + \frac{p^2+6q}{rq}$$

$$M(t) = \left(\frac{p}{1-qe^t}\right)^r \qquad \phi(t) = \left(\frac{p}{1-qe^{it}}\right)^r \qquad P(t) = \left(\frac{p}{1-qt}\right)^r$$

Poisson Distribution

$$p(x) = \frac{e^{-\mu}\mu^x}{x!} \qquad x = 0,1,2,... \qquad \mu > 0$$

$$\mu = \mu \qquad \sigma^2 = \mu \qquad \beta_1 = \frac{1}{\sqrt{\mu}} \qquad \beta_2 = 3 + \frac{1}{\mu}$$

$$M(t) = \exp\left[\mu\left(e^t - 1\right)\right] \qquad \sigma(t) = \exp\left[\mu\left(e^{it} - 1\right)\right] \qquad P(t) = \exp\left[\mu(t - 1)\right]$$

Rectangular (Discrete Uniform) Distribution

$$p(x) = 1/n \qquad x = 1,2,...,n \qquad n \in N$$

$$\mu = \frac{n+1}{2} \qquad \sigma^2 = \frac{n^2 - 1}{12} \qquad \beta_1 = 0 \qquad \beta_2 = \frac{3}{5}\left(3 - \frac{4}{n^2 - 1}\right)$$

$$M(t) = \frac{e^t\left(1 - e^{nt}\right)}{n\left(1 - e^t\right)} \qquad \phi(t) = \frac{e^{it}\left(1 - e^{nit}\right)}{n\left(1 - e^{it}\right)} \qquad P(t) = \frac{t\left(1 - t^n\right)}{n(1 - t)}$$

Continuous Distribution Formulas

Probability density function, $f(x)$

Mean, μ

Variance, s^2

Coefficient of skewness, b_1

Coefficient of Kurtosis, b_2

Moment-generating function, $M(t)$

Characteristic function, $f(t)$

Arcsin Distribution

$$f(x) = \frac{1}{\pi\sqrt{x(1 - x)}} \qquad 0 < x < 1$$

$$\mu = \frac{1}{2} \qquad \sigma^2 = \frac{1}{8} \qquad \beta_1 = 0 \qquad \beta_2 \frac{3}{2}$$

Beta Distribution

$$f(x) = \frac{\Gamma(\alpha+\beta)}{\Gamma(\alpha)\Gamma(\beta)} x^{\alpha-1}(1-x)^{\beta-1} \qquad 0 < x < 1 \qquad \alpha, \beta > 0$$

$$\mu = \frac{\alpha}{\alpha+\beta} \quad \sigma^2 = \frac{\alpha\beta}{(\alpha+\beta)^2(\alpha+\beta+1)} \quad \beta_1 = \frac{2(\beta-\alpha)\sqrt{\alpha+\beta+1}}{\sqrt{\alpha\beta}(\alpha+\beta+2)}$$

$$\beta_2 = \frac{3(\alpha+\beta+1)\left[2(\alpha+\beta)^2 + \alpha\beta(\alpha+\beta-6)\right]}{\alpha\beta(\alpha+\beta+2)(\alpha+\beta+3)}$$

Cauchy Distribution

$$f(x) = \frac{1}{b\pi\left(1+\left(\frac{x-a}{b}\right)^2\right)} \qquad -\infty < x < \infty \qquad -\infty < a < \infty \qquad b > 0$$

m, s^2, b_1, b_2, $M(t)$ do not exist.

$$\phi(t) = \exp\left[ait - b|t|\right]$$

Chi Distribution

$$f(x) = \frac{x^{n-1}e^{-x^2/2}}{2^{(n/2)-1}\Gamma(n/2)} \qquad x \ge 0 \qquad n \in N$$

$$\mu = \frac{\Gamma\left(\frac{n+1}{2}\right)}{\Gamma\left(\frac{n}{2}\right)} \qquad \sigma^2 = \frac{\Gamma\left(\frac{n+2}{2}\right)}{\Gamma\left(\frac{n}{2}\right)} - \left[\frac{\Gamma\left(\frac{n+1}{2}\right)}{\Gamma\left(\frac{n}{2}\right)}\right]^2$$

Chi-Square Distribution

$$f(x) = \frac{e^{-x/2}x^{(\nu/2)-1}}{2^{\nu/2}\Gamma(\nu/2)} \qquad x \ge 0 \qquad \nu \in N$$

$$\mu = \nu \quad \sigma^2 = 2\nu \quad \beta_1 = 2\sqrt{2/\nu} \quad \beta_2 = 3 + \frac{12}{\nu} \quad M(t) = (1-2t)^{-\nu/2},\ t < \frac{1}{2}$$

$$\phi(t) = (1-2it)^{-\nu/2}$$

Erlang Distribution

$$f(x)=\frac{1}{\beta^{n(n-1)!}}x^{n-1}e^{-x/\beta} \qquad x\geq 0 \qquad \beta>0 \qquad n\in N$$

$$\mu=n\beta \qquad \sigma^2=n\beta^2 \qquad \beta_1=\frac{2}{\sqrt{n}} \qquad \beta_2=3+\frac{6}{n}$$

$$M(t)=(1-\beta t)^{-n} \qquad \phi(t)=(1-\beta it)^{-n}$$

Exponential Distribution

$$f(x)=\lambda e^{-\lambda x} \qquad x\geq 0 \qquad \lambda>0$$

$$\mu=\frac{1}{\lambda} \qquad \sigma^2=\frac{1}{\lambda^2} \qquad \beta_1=2 \qquad \beta_2=9 \qquad M(t)=\frac{\lambda}{\lambda-t}$$

$$\phi(t)=\frac{\lambda}{\lambda-it}$$

Extreme-Value Distribution

$$f(x)=\exp\left[-e^{-(x-\alpha)/\beta}\right] \qquad -\infty<x<\infty \qquad -\infty<\alpha<\infty \qquad \beta>0$$

$\mu=\alpha+\gamma\beta$, $\gamma \doteq .5772\ldots$ is Euler's constant $\sigma^2=\frac{\pi^2\beta^2}{6}$, where b_1 is 1.29857 and b_2 is 5.4.

$$M(t)=e^{\alpha t}\Gamma(1-\beta t), \quad t<\frac{1}{\beta} \qquad \phi(t)=e^{\alpha it}\Gamma(1-\beta it)$$

F Distribution

$$f(x)\frac{\Gamma[(\nu_1+\nu_2)/2]\nu_1^{\nu_1/2}\nu_2^{\nu_2/2}}{\Gamma(\nu_1/2)\Gamma(\nu_2/2)}x^{(\nu_1/2)-1}(\nu_2+\nu_1 x)^{-(\nu_1+\nu_2)/2}$$

$$x>0 \qquad \nu_1,\ \nu_2\in N$$

$$\mu=\frac{\nu_2}{\nu_2-2},\ \nu_2\geq 3 \qquad \sigma^2=\frac{2\nu_2^2(\nu_1+\nu_2-2)}{\nu_1(\nu_2-2)^2(\nu_2-4)}, \quad \nu_2\geq 5$$

$$\beta_1 = \frac{(2\nu_1 + \nu_2 - 2)\sqrt{8(\nu_2 - 4)}}{\sqrt{\nu_1}\,(\nu_2 - 6)\sqrt{\nu_1 + \nu_2 - 2}}, \quad \nu_2 \geq 7$$

$$\beta_2 = 3 + \frac{12\left[(\nu_2 - 2)^2(\nu_2 - 4) + \nu_1(\nu_1 + \nu_2 - 2)(5\nu_2 - 22)\right]}{\nu_1(\nu_2 - 6)(\nu_2 - 8)(\nu_1 + \nu_2 - 2)}, \quad \nu_2 \geq 9$$

$M(t)$ does not exist.

$$\phi\left(\frac{\nu_1}{\nu_2}t\right) = \frac{G(\nu_1, \nu_2, t)}{B(\nu_1/2, \nu_2/2)}$$

$B(a,b)$ is the Beta function. G is defined by

$$(m+n-2)G(m,n,t) = (m-2)G(m-2,n,t) + 2itG(m,n-2,t), \quad m,n > 2$$

$$mG(m,n,t) = (n-2)G(m+2,n-2,t) - 2itG(m+2,n-4,t), \quad n > 4$$

$$nG(2,n,t) = 2 + 2itG(2,n-2,t), \quad n > 2$$

Gamma Distribution

$$f(x) = \frac{1}{\beta^{\alpha}\Gamma(\alpha)} x^{\alpha-1}e^{-x/\beta} \qquad x \geq 0 \qquad \alpha, \beta > 0$$

$$\mu = \alpha\beta \qquad \sigma^2 = \alpha\beta^2 \qquad \beta_1 = \frac{2}{\sqrt{\alpha}} \qquad \beta_2 = 3\left(1 + \frac{2}{\alpha}\right)$$

$$M(t) = (1 - \beta t)^{-\alpha} \qquad \phi(t) = (1 - \beta it)^{-\alpha}$$

Half-Normal Distribution

$$f(x) = \frac{2\theta}{\pi}\exp\left[-\left(\theta^2 x^2/\pi\right)\right] \qquad x \geq 0 \qquad \theta > 0$$

$$\mu = \frac{1}{\theta} \qquad \sigma^2 = \left(\frac{\pi - 2}{2}\right)\frac{1}{\theta^2} \qquad \beta_1 = \frac{4 - \pi}{\theta^3} \qquad \beta_2 = \frac{3\pi^2 - 4\pi - 12}{4\theta^4}$$

LaPlace (Double Exponential) Distribution

$$f(x) = \frac{1}{2\beta}\exp\left[-\frac{|x - \alpha|}{\beta}\right] \qquad -\infty < x < \infty \qquad -\infty < \alpha < \infty \qquad \beta > 0$$

$m = a; \quad s^2 = 2b^2; \quad b_1 = 0; \quad b_2 = 6$

$$M(t) = \frac{e^{\alpha t}}{1-\beta^2 t^2} \qquad \phi(t) = \frac{e^{\alpha i t}}{1+\beta^2 t^2}$$

Logistic Distribution

$$f(x) = \frac{\exp[(x-\alpha)/\beta]}{\beta(1+\exp[(x-\alpha)/\beta])^2}$$

$$-\infty < x < \infty; \quad -\infty < \alpha < \infty; \quad -\infty < \beta < \infty$$

$$\mu = \alpha \qquad \sigma^2 = \frac{\beta^2\pi^2}{3} \qquad \beta_1 = 0 \qquad \beta_2 = 4.2$$

$$M(t) = e^{\alpha t}\pi\beta t \csc(\pi\beta t) \qquad \phi(t) = e^{\alpha i t}\pi\beta i t \csc(\pi\beta i t)$$

Lognormal Distribution

$$f(x) = \frac{1}{\sqrt{2\pi\sigma x}} \exp\left[-\frac{1}{2\sigma^2}(1nx-\mu)^2\right]$$

$$x > 0; \quad -\infty < \mu < \infty; \quad \sigma > 0$$

$$\mu = e^{\mu+\sigma^2/2} \qquad \sigma^2 = e^{2\mu+\sigma^2}\left(e^{\sigma^2}-1\right)$$

$$\beta_1 = \left(e^{\sigma^2}+2\right)\left(e^{\sigma^2}-1\right)^{1/2} \qquad \beta_2 = \left(e^{\sigma^2}\right)^4 + 2\left(e^{\sigma^2}\right)^3 + 3\left(e^{\sigma^2}\right)^2 - 3$$

Noncentral Chi-Square Distribution

$$f(x) = \frac{\exp\left[-\frac{1}{2}(x+\lambda)\right]}{2^{\nu/2}} \sum_{j=0}^{\infty} \frac{x^{(\nu/2)+j-1\lambda j}}{\Gamma\left(\frac{\nu}{2}+j\right)2^{2j}j!}$$

$$x > 0; \quad \lambda > 0; \quad \nu \in N$$

$$\mu = \nu + \lambda \qquad \sigma^2 = 2(\nu+2\lambda) \qquad \beta_1 = \frac{\sqrt{8}(\nu+3\lambda)}{(\nu+2\lambda)^{3/2}} \qquad \beta_2 = 3 + \frac{12(\nu+4\lambda)}{(\nu+2\lambda)^2}$$

$$M(t) = (1-2t)^{-\nu/2}\exp\left[\frac{\lambda t}{1-2t}\right] \qquad \phi(t) = (1-2it)^{-\nu/2}\exp\left[\frac{\lambda i t}{1-2it}\right]$$

Noncentral F Distribution

$$f(x)=\sum_{i=0}^{\infty}\frac{\Gamma\left(\frac{2i+\nu_1+\nu_2}{2}\right)\left(\frac{\nu_1}{\nu_2}\right)^{(2i+\nu_1)/2}x^{(2i+\nu_1-2)/2}e^{-\lambda/2}\left(\frac{\lambda}{2}\right)}{\Gamma\left(\frac{\nu_2}{2}\right)\Gamma\left(\frac{2i+\nu_1}{2}\right)\nu_1!\left(1+\frac{\nu_1}{\nu_2}x\right)^{(2i+\nu_1+\nu_2)/2}}$$

$$x>0 \qquad \nu_1,\nu_2\in N \qquad \lambda>0$$

$$\mu=\frac{(\nu_1+\lambda)\nu_2}{(\nu_2-2)\nu_1}, \quad \nu_2>2$$

$$\sigma^2=\frac{(\nu_1+\lambda)^2+2(\nu_1+\lambda)\nu_2^2}{(\nu_2-2)(\nu_2-4)v_1^2}-\frac{(\nu_1+\lambda)^2\nu_2^2}{(\nu_2-2)^2\nu_1^2}, \quad \nu_2>4$$

Noncentral t Distribution

$$f(x)=\frac{\nu^{\nu/2}}{\Gamma\left(\frac{\nu}{2}\right)}\frac{e^{-\delta^2/2}}{\sqrt{\pi}\left(\nu+x^2\right)^{(\nu+1)/2}}\sum_{i=0}^{\infty}\Gamma\left(\frac{\nu+i+1}{2}\right)\left(\frac{\delta^i}{i!}\right)\left(\frac{2x^2}{\nu+x^2}\right)^{i/2}$$

$$-\infty<x<\infty; \quad -\infty<\delta<\infty; \quad \nu\in N$$

$$\mu_r'=c_r\frac{\Gamma\left(\frac{\nu-r}{2}\right)\nu^{r/2}}{2^{r/2}\Gamma\left(\frac{\nu}{2}\right)}, \quad \nu>r, \qquad c_{2r-1}=\sum_{i=1}^{r}\frac{(2r-1)!\delta^{2r-1}}{(2i-1)!(r-i)!2^{r-i}},$$

$$c_{2r}=\sum_{i=0}^{r}\frac{(2r)!\delta^{2i}}{(2i)!(r-i)!2^{r-i}}, \qquad r=1,2,3,\ldots$$

Normal Distribution

$$f(x)=\frac{1}{\sigma\sqrt{2\pi}}\exp\left[-\frac{(x-\mu)^2}{2\sigma^2}\right]$$

$$-\infty<x<\infty; \quad -\infty<\mu<\infty; \quad \sigma>0$$

$$\mu=\mu \qquad \sigma^2=\sigma^2 \qquad \beta_1=0 \qquad \beta_2=3 \qquad M(t)=\exp\left[\mu t+\frac{t^2\sigma^2}{2}\right]$$

$$\phi(t)=\exp\left[\mu it-\frac{t^2\sigma^2}{2}\right]$$

Pareto Distribution

$$f(x) = \theta a^{\theta} / x^{\theta+1} \qquad x \geq a \qquad \theta > 0 \qquad a > 0$$

$$\mu = \frac{\theta a}{\theta - 1}, \quad \theta > 1 \qquad \sigma^2 = \frac{\theta a^2}{(\theta-1)^2(\theta-2)}, \qquad \theta > 2$$

$M(t)$ does not exist.

Rayleigh Distribution

$$f(x) = \frac{x}{\sigma^2}\exp\left[-\frac{x^2}{2\sigma^2}\right] \qquad x \geq 0 \qquad \sigma = 0$$

$$\mu = \sigma\sqrt{\pi/2} \qquad \sigma^2 = 2\sigma^2\left(1 - \frac{\pi}{4}\right) \qquad \beta_1 = \frac{\sqrt{\pi}}{4}\frac{(\pi-3)}{\left(1-\frac{\pi}{4}\right)^{3/2}}$$

$$\beta_2 = \frac{2 - \frac{3}{16}\pi^2}{\left(1 - \frac{\pi}{4}\right)^2}$$

t Distribution

$$f(x) = \frac{1}{\sqrt{\pi\nu}}\frac{\Gamma\left(\frac{\nu+1}{2}\right)}{\Gamma\frac{\nu}{2}}\left(1 + \frac{x^2}{\nu}\right)^{-(\nu+1)/2} \qquad -\infty < x < \infty \qquad \nu \in N$$

$$\mu = 0, \quad \nu \geq 2 \qquad \sigma^2 = \frac{\nu}{\nu - 2}, \quad \nu \geq 3 \qquad \beta_1 = 0, \quad \nu \geq 4$$

$$\beta_2 = 3 + \frac{6}{\nu - 4}, \quad \nu \geq 5$$

M(t) does not exist.

$$\phi(t) = \frac{\sqrt{\pi}\Gamma\left(\frac{\nu}{2}\right)}{\Gamma\left(\frac{\nu+1}{2}\right)}\int_{-\infty}^{\infty}\frac{e^{itz\sqrt{\nu}}}{\left(1+z^2\right)^{(\nu+1)/2}}dz$$

Triangular Distribution

$$f(x)=\begin{cases}0 & x\le a\\ 4(x-a)/(b-a)^2 & a<x\le(a+b)/2\\ 4(b-x)/(b-a)^2 & (a+b)/2<x<b\\ 0 & x\ge b\end{cases}$$

$$-\infty<a<b<\infty$$

$$\mu=\frac{a+b}{2}\qquad \sigma^2=\frac{(b-a)^2}{24}\qquad \beta_1=0\qquad \beta_2=\frac{12}{5}$$

$$M(t)=-\frac{4\left(e^{at/2}-e^{bt/2}\right)^2}{t^2(b-a)^2}\qquad \phi(t)=\frac{4\left(e^{ait/2}-e^{bit/2}\right)^2}{t^2(b-a)^2}$$

Uniform Distribution

$$f(x)=\frac{1}{b-a}\qquad a\le x\le b\qquad -\infty<a<b<\infty$$

$$\mu=\frac{a+b}{2}\qquad \sigma^2=\frac{(b-a)^2}{12}\qquad \beta_1=0\qquad \beta_2=\frac{9}{5}$$

$$M(t)=\frac{e^{bt}-e^{at}}{(b-a)t}\qquad \phi(t)=\frac{e^{bit}-e^{ait}}{(b-a)it}$$

Weibull Distribution

$$f(x)=\frac{\alpha}{\beta^\alpha}x^{\alpha-1}e^{-(x/\beta)^\alpha}\qquad x\ge 0\qquad \alpha,\beta>0$$

$$\mu=\beta\Gamma\left(1+\frac{1}{\alpha}\right)\qquad \sigma^2=\beta^2\left[\Gamma\left(1+\frac{2}{\alpha}\right)-\Gamma^2\left(1+\frac{1}{\alpha}\right)\right]$$

$$\beta_1=\frac{\Gamma\left(1+\frac{3}{\alpha}\right)-3\Gamma\left(1+\frac{1}{\alpha}\right)\Gamma\left(1+\frac{2}{\alpha}\right)+2\Gamma^3\left(1+\frac{1}{\alpha}\right)}{\left[\Gamma\left(1+\frac{2}{\alpha}\right)-\Gamma^2\left(1+\frac{1}{\alpha}\right)\right]^{3/2}}$$

$$\beta_2=\frac{\Gamma\left(1+\frac{4}{\alpha}\right)-4\Gamma\left(1+\frac{1}{\alpha}\right)\Gamma\left(1+\frac{3}{\alpha}\right)+6\Gamma^2\left(1+\frac{1}{\alpha}\right)\Gamma\left(1+\frac{2}{\alpha}\right)-3\Gamma^4\left(1+\frac{1}{\alpha}\right)}{\left[\Gamma\left(1+\frac{2}{\alpha}\right)-\Gamma^2\left(1+\frac{1}{\alpha}\right)\right]^2}$$

Variate Generation Techniques

From Leemis, L. M. 1987. Variate Generation for Accelerated Life and Proportional Hazards Models, *Operations Research* 35(6).

Notation

Let $h(t)$ and $H(t) = \int_0^t h(\tau)d\tau$ be the hazard and cumulative hazard functions, respectively, for a continuous nonnegative random variable *T*, the lifetime of the item under study. The $q \times 1$ vector *z* contains covariates associated with a particular item or individual. The covariates are linked to the lifetime by the function *Y*(*z*), which satisfies *Y*(0 = 1) and *Y*(*z*) ≥ 0 for all *z*. A popular choice is $\Psi(z) = e^{\beta' z}$, where β is a vector of regression coefficients.

The cumulative hazard function for T in the *accelerated life* model is

$$H(t) = H_0(t\,\Psi\,(z)),$$

where H_0 is a baseline cumulative hazard function. Note that when $z = 0$, $H_0 \equiv H$. In this model, the covariates accelerate (Y(*z*) > 1) or decelerate (Y(*z*) < 1), the rate at which the item moves through time. The *proportional* hazards model

$$H(t) = \Psi\,(z) H_0\,(t)$$

increases (Y(*z*) > 1) or decreases (Y(*z*) < 1) the failure rate of the item by factor Y(*z*) for all values of *t*.

Variate Generation Algorithms

The literature shows that the cumulative hazard function, *H*(*T*), has a unit exponential distribution. Therefore, a random variate *t* corresponding to a cumulative hazard function *H*(*t*) can be generated by

$$t = H^{-1}\left(-\log(u)\right),$$

where *u* is uniformly distributed between 0 and 1. In the accelerated life model, since time is being expanded or contracted by a factor Y(*z*), variates are generated by

$$t = \frac{H_0^{-1}(-\log(u))}{\Psi(z)}$$

In the proportional hazards model, equating –log(*u*) to *H*(*t*) yields the variate generation formula

$$t = H_0^{-1}\left(\frac{-\log(u)}{\Psi(z)}\right).$$

TABLE A.2

Formulas for Generating Event Times from a Renewal or Nonhomogeneous Poisson Process

	Renewal	NHPP
Accelerated life	$t = a + \frac{H_0^{-1}(-\log(u))}{\Psi(z)}$	$t = \frac{H_0^{-1}(H_0(a\Psi(z)) - \log(u))}{\Psi(z)}$
Proportional hazards	$t = a + H_0^{-1}\left(\frac{-\log(u)}{\Psi(z)}\right)$	$t = H_0^{-1}\left(H_0(a) - \frac{\log(u)}{\Psi(z)}\right)$

Table A.2 shows formulas for generating event times from a renewal or nonhomogeneous Poisson process. In addition to generating individual lifetimes, these variate generation techniques may also be applied to point processes. A renewal process, for example, with time between events having a cumulative hazard function *H*(*t*), can be simulated by using the appropriate generation formula for the two cases just shown. These variate generation formulas must be modified, however, to generate variates from a nonhomogeneous Poisson process (NHPP).

In a NHPP, the hazard function, h(t), is equivalent to the intensity function, which governs the rate at which events occur. To determine the appropriate method for generating values from an NHPP, assume that the last event in a point process has occurred at time a. The cumulative hazard function for the time of the next event conditioned on survival to time a is

$$H_{T|T>a}(t) = H(t) - H(a) \quad t > a$$

In the accelerate life model, where $H(t) = H_0(t\Psi(z))$, the time of the next event is generated by

$$t = \frac{H_0^{-1}(H_0(a\Psi(z)) - \log(u))}{\Psi(z)}$$

If we equate the conditional cumulative hazard function to –log(*u*), the time of the next event in the proportional hazards case is generated by

$$t = H_0^{-1}\left(H_0(a) - \frac{\log(u)}{\Psi(z)}\right).$$

Example

The exponential power distribution is a flexible two-parameter distribution with cumulative hazard function

$$H(t) = e^{(t/\alpha)^\gamma} - 1 \quad \alpha > 0, \quad \gamma > 0, \quad t > 0$$

and inverse cumulate hazard function

$$H^{-1}(y) = \alpha\left[\log(y+1)\right]^{1/\gamma}.$$

Assume that the covariates are linked to survival by the function $\Psi(z) = e^{\beta'z}$ in the accelerated life model. If an NHPP is to be simulated, the baseline hazard function has the exponential power distribution with parameters α and γ, and the previous event has occurred at time a, then the next event is generated at time

$$t = \alpha e^{-\beta'z}\left[\log\left(e^{(ae\beta'z/\alpha)^{\gamma}} - \log(u)\right)\right]^{1/\gamma},$$

where u is uniformly distributed between 0 and 1.

Index